Mysteries in Muscle Contraction

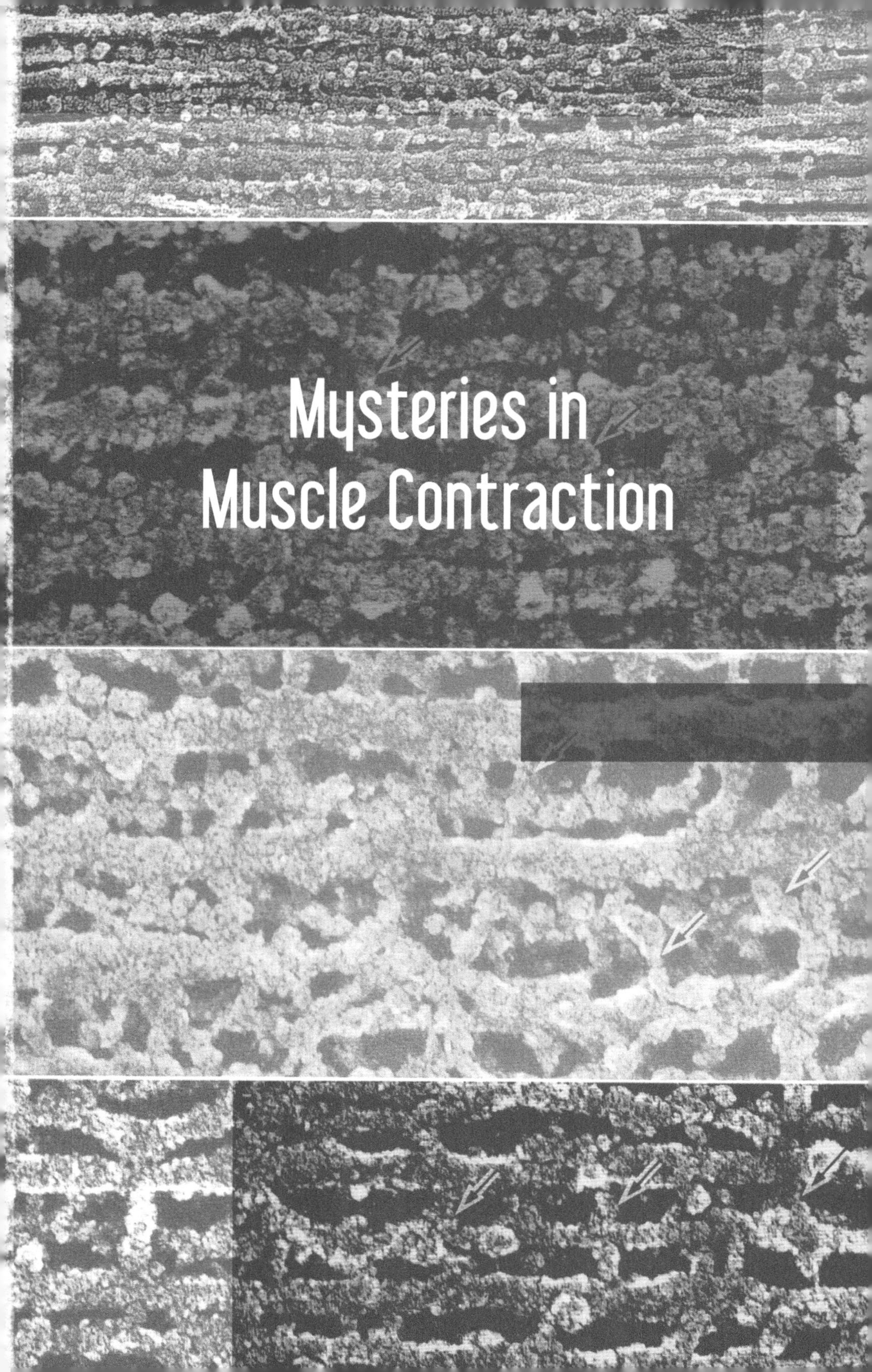

Mysteries in Muscle Contraction

Evidence against Current Dogmas

Haruo Sugi

PAN STANFORD PUBLISHING

Published by

Pan Stanford Publishing Pte. Ltd.
Penthouse Level, Suntec Tower 3
8 Temasek Boulevard
Singapore 038988

Email: editorial@panstanford.com
Web: www.panstanford.com

British Library Cataloguing-in-Publication Data
A catalogue record for this book is available from the British Library.

Mysteries in Muscle Contraction: Evidence against Current Dogmas

ISBN 978-981-4774-48-2 (Hardcover)
ISBN 978-1-315-10191-0 (eBook)

Printed in the USA

Contents

Preface xv

1. Historical Background 1

1.1 Structural Organization of Striated Muscle 2
1.2 Discovery of the Sliding Filament Mechanism 3
1.3 Myofilament Structure and Possible Role of Myosin Heads in Producing Myofilament Sliding 5
1.4 A. F. Huxley's Contraction Model 8

2. Dogmas of Huxley–Simmons Model Not Supported Experimentally 13

2.1 Huxley–Simmons Contraction Model: Dogmas about Myofilament Rigidity and Myosin Head Tilting 13
2.2 Myofilaments Are Not Completely Rigid 16
2.2.1 Electron Microscopic Studies on the Extensibility of Myofilaments 16
2.2.2 Electron Microscopic Studies on the Extensibility of Actin Filament Extensibility Using Freeze-Fracture Method 18
2.3 Myosin Head Tilting Was Not Supported Experimentally 20
2.4 Quick Force Recovery Following Quick Release Is a Complex Viscoelastic Phenomenon 21

3. New Methods and New Dogmas and Their Shortcomings 27

3.1 The Lymn–Taylor Scheme of Actomyosin ATPase Reaction 27

3.2 Electron Microscopic Studies on the Static Structure of Rigor Actin–Myosin Complex 31
3.3 Development of in vitro Motility Assay Systems 33
3.3.1 Fluorescent Actin Filament versus Myosin System 33
3.3.2 Actin Cable versus Myosin System 34
3.4 The Optical Trap System 38
3.4.1 Advantages of the Optical Trap System 38
3.4.2 Shortcomings of the Optical Trap System 40

4. The Swinging Lever Arm Mechanism: A Most Prominent Current Dogma 43

4.1 Construction of the Swinging Lever Arm Hypothesis 43
4.2 Shortcomings of the Swinging Lever Arm Hypothesis 45

5. Early Electron Microscopic Studies on Myosin Head Movement Using Quick-Freeze, Deep-Etch Replica Method 49

5.1 General Feature of the Freeze-Etch Replica Images 50
5.2 Myosin Head Angle Distribution in Freeze-Etch Replica Images 53
5.2.1 Myosin Head Angular Distribution in Relaxed and Contracting States 56
5.2.2 Myosin Head Angular Distribution in Rigor States 57

6. Electron Microscopic Recording of ATP-Induced Myosin Head Movement Using the Gas Environmental Chamber 61

6.1 The Gas Environmental Chamber 62
6.1.1 Carbon Sealing Film 63
6.1.2 Iontophoretic Application of ATP 64
6.1.3 Critical Electron Dose Not to Impair Contractile Function of Myofilaments 65
6.1.4 Position Marking of Individual Myosin Heads 66
6.2 Preliminary Experiments on Myosin–Paramyosin Core Complex in the Absence of Actin Filaments 68

6.2.1 Observation and Recording of Filament Image 68
6.2.2 Data Analysis 69
6.2.3 Stability in Time-Averaged Position of Myosin Heads 70
6.2.4 ATP-Induced Myosin Head Movement 72
6.3 Experiments on Synthetic Vertebrate Muscle Myosin Filaments in the Absence of Actin Filaments 74
6.3.1 Preparation of Vertebrate Muscle Bipolar Myosin Filaments 74
6.3.2 Stability in the Time-Averaged Position of Myosin Heads in the Absence of ATP 75
6.3.3 ATP-Induced Unidirectional Myosin Head Movement in the Absence of Actin Filaments 77
6.3.4 Reversal in the Direction of ATP-Induced Myosin Head Movement Across the Bare Region 78
6.3.5 Reversibility of ATP-Induced Myosin Head Movement 80
6.4 Amplitude of Myosin Head Movement at Three Different Regions of Myosin Head 82

7. Evidence against Current Dogmas and Mechanism of Muscle Contraction as Revealed by the Effect of Antibodies on Myosin Head 85

7.1 Antibody 1 Has No Effect on Both Muscle Contraction and in vitro Actin–Myosin Sliding 85
7.2 Antibody 2 Inhibits in vitro Actin–Myosin Sliding but Has No Effect on Muscle Contraction 87
7.3 Antibody 3 Has No Significant Effect on in vitro Actin–Myosin Sliding but Inhibits Isometric Force Development in Muscle Without Changing ATPase Activity 90
7.4 Interpretation of the Effect of Antibodies on Muscle Contraction and in vitro Actin–Myosin Sliding 94
7.4.1 Diffusion of Antibody (IgG) Molecules into Skinned Muscle Fibers 94

7.4.2 Ineffectiveness of Antibody 1 on Both Muscle Contraction and in vitro Actin–Myosin Sliding 95
7.4.3 Marked Inhibitory Effect of Antibody 2 on in vitro Actin–Myosin Sliding 97
7.4.4 Inhibitory Effect of Antibody 3 and Anti-S-2 Antibody on Muscle Contraction 98
7.4.5 Evidence for Functional Communication between Myosin Head CAD and Myosin S-2 Regions 100
7.4.6 Regulation of Myosin Head-Actin Binding Strength by Myosin Head LD and Myosin S-2 102
7.4.7 Evidence for Cooperativity of Myosin Two Heads in Force Generation in Muscle 105
7.4.8 Fiber Stiffness and Internal Resistance 109
7.4.9 Effect of p-PDM Modification on Myofilament Lattice Structure of Skinned Muscle Fibers 110

8. Electron Microscopic Recording of Myosin Head Power Stroke: Evidence That Myosin Heads Take the Neutral Position When Detached from Actin Filaments 113

8.1 Preparation of Actin and Myosin Filament Mixture 114
8.2 Amplitude of ATP-Induced Myosin Head Power Stroke 115
8.3 Reversibility of ATP-Induced Myosin Head Power Stroke 118
8.4 Enhanced Performance of Myosin Heads at Low Ionic Strength: Agreement of Results between EC and Physiological Experiments 119
8.4.1 Physiological Experiments 119
8.4.2 EC Experiments 121
8.5 Two Different Modes of Myosin Head Power Stroke As Revealed by EC Experiments 124

8.6 Sequence of Changes in Myosin Head Configuration during ATP-Induced Myosin Head Power and Recovery Strokes 125
8.6.1 Myosin Heads Take Neutral Position Where They Exert No Force When Detached from Actin Filaments 126
8.6.2 Why Only Two Definite Myosin Head Positions Are Recorded in EC Experiments 128
8.6.2.1 ATP-induced myosin head recovery stroke in the absence of actin filaments 129
8.6.2.2 ATP-induced myosin head power stroke in the filament mixture 130
8.7 Possible Electrostatic Interaction between Myosin Head and Actin Filaments 132
8.7.1 Anisotropic Nature of Contracting Muscle as Revealed by the Use of Ultrasonic Waves 132
8.7.2 Other Evidence for the Electrostatic Actin-Myosin Interaction 135

9. The Fenn Effect: A Fundamental Mystery in Muscle 139

9.1 Historical Discovery of the Fenn Effect 139
9.2 Hill's Force–Velocity Equation 141
9.3 Force-Velocity Relation Obtained from Experiments with Single Muscle Fibers 143
9.3.1 Advantages of Single Muscle Fibers over Whole Muscles 143
9.3.2 Force–Velocity Relation Obtained from Single Intact and Skinned Muscle Fibers 144
9.3.3 Isotonic Velocity Transients Preceding Constant Velocity Shortening in Single Intact Muscle Fibers 145
9.4 The Fenn Effect Is Observed in Both Myofibrils and in vitro Motility Assay System 148
9.4.1 Experiments on Muscle Myofibrils 149

9.4.2 Experiments on in vitro Motility Assay System 152

10. Load-Dependent Mechanical Efficiency of Myosin Head Performance 155

10.1 Classical Experiments on Mechanical Efficiency of Whole Muscles 156

10.2 Measurement of Mechanical Efficiency of Individual Myosin Heads during Auxotonic Shortening 157

10.2.1 Experimental Design 157

10.2.2 Features of Auxotonic Shortening of Laser Flash-Activated Muscle Fibers 158

10.2.3 Load-Dependent Mechanical Efficiency of Myosin Heads during Auxotonic Shortening 161

10.3 Measurement of Mechanical Efficiency of Individual Myosin Heads during Isotonic Shortening 165

10.3.1 Features of Power Output in the Early Phase of Isotonic Shortening 165

10.3.2 Mechanical Efficiency of Myosin Heads during Afterloaded Isotonic Shortening 167

10.3.3 Dependence of Mechanical Efficiency of Myosin Heads on Isotonic Load 169

11. Past History-Dependence of Myosin Head Performance 171

11.1 Force Enhancement after Slow Stretch and Force Deficit after Slow Release 171

11.2 Heat Production of Whole Muscle during Slow Stretch 172

11.3 Enhancement of Mechanical Performance in Stretched Muscle Fibers: Possible Storage of Energy by Stretch 174

11.4 Measurement of Muscle Fiber Stiffness in Single Tetanized Muscle Fibers before, during and after Slow Stretches 176

11.5 Time-Resolved X-Ray Diffraction Studies on Equatorial Reflections from Contracting Muscle during Slow Stretch 178
11.6 Electron Microscopic Studies on Stretch-Induced Disorder of Myofilament Lattice 182
11.7 Measurement of Muscle Fiber Stiffness in Single Tetanized Muscle Fibers before, during, and after Slow Release 184
11.8 Longitudinal Stability of Sarcomere Length along the Fiber 186

12. Reminiscences of My Early Work: I. Summation of Force Responses in Muscle 189

12.1 A. V. Hill's Concept of Active State: A Wrong Dogma 190
12.2 Muscle Contraction Is Produced by Depolarization of Surface Membrane 191
12.3 Transverse Tubular System Transmitting Surface Membrane Depolarization into the Fiber 192
12.4 Summation of Isometric Twitch: Argument about the Baseline from Which the Second Twitch Starts 193
12.5 Summation of Force Response in Crayfish Muscle Fibers 194
12.5.1 Observations on Vertically Suspended Muscle Fibers 194
12.5.2 Summation of Force Response by Two Successive Transversely Applied Current Pulses 196
12.5.2.1 Two successive current pulses of opposite directions 196
12.5.2.2 Two successive current pulses of the same direction 199

13. Reminiscences of My Early Work: II. Transverse Spread of Contraction Initiated by Local Membrane Depolarization in Single Muscle Fibers 203

13.1 Crayfish Muscle Fibers 203

13.1.1 General Features of Local Membrane Depolarization 203
13.1.2 Inward Spread of Contraction Initiated by Moderate Membrane Depolarization 205
13.1.3 Summation of Local Contraction Initiated by Moderate Membrane Depolarization 206
13.1.4 Transverse Spread of Local Contraction Initiated by Strong Current 208
13.1.5 Measurement of Velocity of Transverse Spread of Contraction 210
13.1.6 Two Different Types of Spread of Contraction 211
13.2 Frog Muscle Fibers 212
13.2.1 General Features of Local Membrane Depolarization 212
13.2.2 Inward Spread of Contraction Initiated by Moderate Membrane Depolarization 213
13.2.3 Transverse Spread of Contraction Initiated by Strong Current 214
13.2.4 Measurement of Velocity of Transverse Spread of Contraction 216
13.3 Accommodation of E–C Coupling Process as Studied by Local Membrane Depolarization with Linearly Rising Currents 218
13.3.1 Membrane Potential Changes during Application of Linearly Rising Currents for Local Activation 219
13.3.2 Accommodation Curves of Excitation–Contraction Coupling 220
13.3.3 Time Course of Development and Decay of Accommodation Process 222
13.3.4 Accommodation of Action Potential in Muscle Fiber 224
13.4 Inward and Transverse Spread of Contraction Studied Using High-Speed Cinematography 224
13.4.1 General Features of Phasic-Type Local Contraction Initiated by Moderate Membrane Depolarization 225

13.4.2 Effect of Increase in Area of Depolarization 228
13.4.3 Effect of Magnitude of Depolarization 230
13.4.4 Transverse Spread of Contraction Across the Center to the Opposite Side of the Fiber 231
13.4.5 Effect of Temperature 233
13.4.6 Shortening Velocity of Sarcomeres 234
13.4.7 Graded-Type Local Contraction 235
13.4.8 Possible Presence and Role of Regenerative Process in Muscle Contraction 23

14. Epilogue **241**

References 249
Index 261

Preface

A most characteristic function of animals is their ability to move by muscle. This volume gives readers an overview of the current knowledge about the molecular mechanism of muscle contraction but definitely differs from any published guidebooks in the following respects.

First, it provides general readers a comprehensive overview of the experimental work, hitherto made on the structure and function of skeletal muscle. As described in Chapter 1, cross-striation of skeletal muscle was a complete mystery for investigators until the monumental discovery of sliding filament mechanism by Hugh E. Huxley and Jean Hanson in 1954, which showed that anatomical units of cross-striation, i.e., sarcomeres, are also a functional unit of muscle. The history of natural science indicates that a new field of research work is created by a breakthrough work achieved by pioneers or giants and is followed by a period of "normal science," in which many investigators gather in the new research field to solve remaining mysteries. Although more than 50 years have passed since the discovery of the sliding filament mechanism, the muscle research field is still in the period of normal science, despite the fact that, according to many current guidebooks and textbooks, mysteries about the sliding filament mechanism have already been solved. Chapters 2–4 deal with how dogmas to explain myofilament sliding coupled with ATP hydrolysis were developed and denied with the progress of experimental methods.

Second, this volume frankly criticizes current dogmas on the mechanism of muscle contraction, which frequently appear in a number of guidebooks and textbooks. During my research work over more than half a century, I became aware that mainly due to severe competition among muscle investigators for promotion and for scientific grant, some prominent investigators, who developed dogmas constructed from their experimental results, tend to ignore or destroy the publications that are not consistent with their dogmas. Such behavior of prominent investigators retards the progress of the research field, which has been opened by pioneers. Reflecting the above-mentioned situation, there are many access-free journals that accept papers that have been unduly rejected by mainstream journals for unknown reasons.

Third, this volume present in Chapters 5–8 the experimental results obtained in my laboratory. Because of fantastic support of Teikyo University, I was able to perform experiments with many bright coworkers using various expensive apparatuses, though my papers are not well cited despite that they have been published in mainstream journals. Especially, I feel very proud that I was able to obtain the electron microscopic recording of ATP-induced myosin head power and recovery strokes in living myosin filament (Chapters 6–8), which were highly evaluated by two giants, Andrew F. Huxley, who made a monumental work on the ionic mechanism of action potential generation (1952) with Alan L. Hodgkin (1952), and Hugh E. Huxley, who discovered sliding filament mechanism in muscle contraction with Jean Hanson (1954), while my work is generally ignored by people constituting the normal science.

Fourth, Chapters 9–11 deal with muscle energetics, i.e., how muscle converts the energy derived from ATP hydrolysis into mechanical work, including its historical aspects as well as our experimental work on the mode performance of individual myosin heads in various conditions made in my laboratory.

Finally, Chapters 12 and 13 concern the reminiscence of my early experimental work when I started my research carrier as a muscle physiologist. Initially, I felt somewhat hesitant to write about it, but as I began to write it, I felt great pleasure recalling that in the mid-1960s, one could perform original work with a simple experimental setup based on one's own idea. In this connection, I would like to express my thanks to the late Professor Koji Uchizono, who generously let me work freely in his laboratory.

I heartily hope that general readers will become interested in muscle research and feel the atmosphere of "live" science in this limited research area. I also hope investigators, especially young ones, will be inspired by this volume to start challenging the remaining mysteries in muscle contraction.

Haruo Sugi
Tokyo, October 2017

Chapter 1

Historical Background

A most prominent characteristic of animals is their ability to move by muscle contraction. The mechanism of contraction has therefore attracted investigators' interest over many years. Although it has been known since the 19th century that the skeletal muscle, which produces the movement of the animal body, is cross-striated and is called striated muscle, the functional meaning of the striation has remained a mystery over many years. Even in the early half of the 1900s, no investigators were aware of the functional significance of cross-striations in the skeletal muscle.

At that time, it was generally believed that the muscle was composed of numerous contractile molecular chains extending along the entire muscle length; in the resting muscle, the molecular chain is in the extended state, and in the contracting muscle, folding of the molecular chain takes place and results in shortening and force generation in the muscle (Buchthal and Kaiser, 1951). This hypothesis was called the transmutation theory and could explain almost all mechanical properties of the resting and the contracting muscle. The transmutation theory was proved wrong, and this tells us that the ability of a theory to explain many known phenomena does not necessarily mean that it is correct.

Mysteries in Muscle Contraction: Evidence against Current Dogmas
Haruo Sugi

ISBN 978-981-4774-48-2 (Hardcover), 978-1-315-10191-0 (eBook)
www.panstanford.com

1.1 Structural Organization of Striated Muscle

Figure 1 illustrates the structural organization of the vertebrate skeletal muscle. A skeletal muscle is firmly connected to the bone at both tendinous ends and is composed of long and parallel muscle fibers (diameter, 50–100 μm), which are multinucleate giant single cells. A muscle fiber is composed of a number of myofibrils (diameter, ~1 μm). Cross-striations are clearly visible in both muscle fibers and myofibrils.

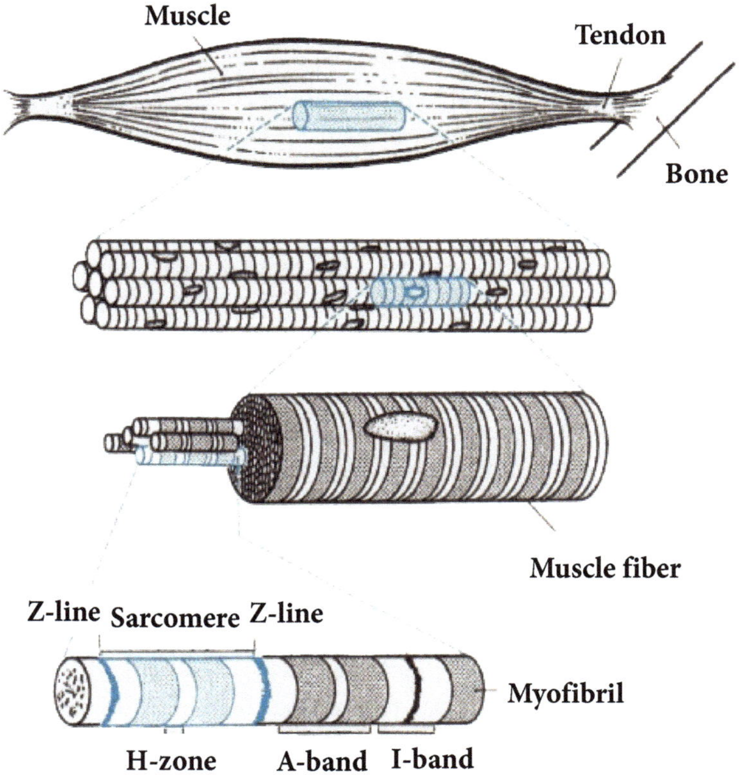

Figure 1 Structural organization of vertebrate skeletal muscle. A muscle consists of muscle fibers, which contain myofibrils. The striation pattern is composed of A- and I-bands, Z-line and H-zone.

In the early 1940s, Szent-Györgyi and coworkers found that the muscle contains two different contractile proteins—actin and myosin—and showed ATP-induced shortening of actomyosin thread, i.e., mixture of actin and myosin, suggesting that muscle contraction results from the interaction between actin and myosin, coupled with ATP hydrolysis (Szent-Györgyi, 1951). These findings were followed by the monumental discovery of the sliding filament mechanism.

1.2 Discovery of the Sliding Filament Mechanism

H. E. Huxley started his research career by studying the X-ray diffraction pattern from the frog skeletal muscle and found a hexagonal double array of filaments (H. E. Huxley, 1953a). When he visited my laboratory in 1978, he told me the following story about his monumental discovery of the sliding filament mechanism.

When he found the hexagonal myofilament lattice in the muscle, Professor Dorothy Hodgkin asked him to visualize his finding. Then he studied the muscle structure electron microscopically and found that the two contractile proteins, actin and myosin, exist in the muscle in the form of two different myofilaments and constitute overlapping double filament arrays (Huxley, 1953b).

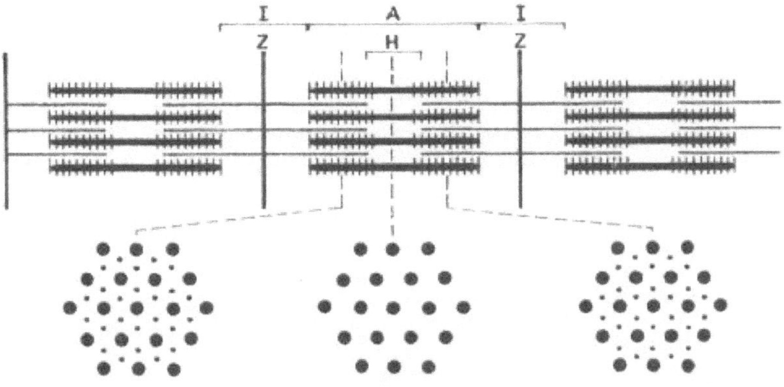

Figure 2 Hexagonal arrangement of thick (myosin) and thin (actin) filaments in vertebrate skeletal muscle. For further explanation, see text.

As can be seen in Fig. 2, the arrangement of actin and myosin filaments corresponded well to the pattern of cross-striation observed under a light microscope, consisting of alternate anisotropic A-band and isotropic I-band with central Z-line. The arrangement of actin and myosin filaments indicates not only the structural basis of cross-striations but also its functional significance.

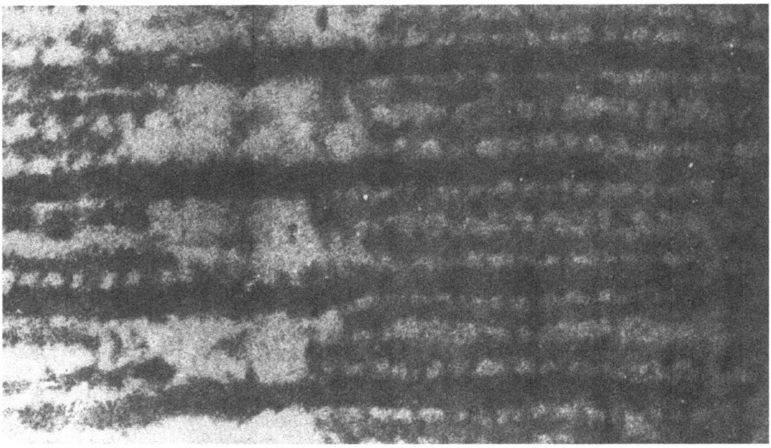

Figure 3 Electron micrograph showing part of A-band where actin and myosin filaments run in parallel with each other. Note cross connections between actin and myosin filaments. From Huxley (1953).

In the longitudinal section of vertebrate skeletal muscle, H. E. Huxley noticed the existence of cross connections between actin and myosin filaments (Fig. 3). When I first invited him to a muscle symposium that I organized, he told me, "*As soon as I found cross connections between the two filaments, an idea occurred to me that the filaments slide past each other by the action of these cross connections.*" Although the cross connections he observed at that time now appear to be an artifact arising from the compression of longitudinal sections due to incomplete techniques to prepare ultrathin sections, it does not affect his deep insight into the mechanism of muscle contraction, which I believe to be an incarnation of human wisdom.

During the course of his pioneering electron microscopic work at the Massachusetts Institute of Technology, the only place in the world where investigators were allowed to use the electron microscope for biological research, H. E. Huxley met J. Hanson, an expert of phase-contrast microscopy, and they started working together productively using myofibrils isolated from the vertebrate skeletal muscle. Their findings (Hanson and Huxley, 1953; Huxley and Hanson, 1954) are summarized as follows: (1) When myofibrils are activated to shorten with ATP, the I-band width decreases, while the A-band width remains unchanged. (2) The magnitude of decrease in the I-band width at both sides of the A-band is equal to the decrease in the width of H-zone, i.e., the less dense region at the center of the A-band. (3) When myosin is extracted from myofibrils with high ionic strength solution, the dense material in the A-band is removed.

These results clearly indicate that (1) the two contractile proteins, actin and myosin, are present in the muscle in the form of two different filaments, i.e., actin and myosin filaments; (2) myosin filaments are located in the A-band, while actin filaments extend in either direction from the Z-line to penetrate in between myosin filaments; and (3) muscle contraction is caused by relative sliding between actin and myosin filaments, in such a way that the amount of overlap between the filaments increases; and (4) the length of actin and myosin filaments remains unchanged during muscle contraction. Thus, the cross-striations are shown to have structural and functional significance. The region between the adjacent Z-band is called the sarcomere and is regarded as the functional unit of the muscle. More exactly, the half-sarcomere is the functional unit, since the sarcomere is symmetrical at both sides of the A-band center, i.e., the H-zone.

1.3 Myofilament Structure and Possible Role of Myosin Heads in Producing Myofilament Sliding

Considerable progress has been made concerning the structure and function of actin and myosin filaments as well as actin and myosin molecules. As shown in Fig. 4A, a myosin molecule (MW 450,000) consists of two pear-shaped heads and a 156 nm-long rod, and can be split into two parts: the 113 nm long rod, known as light meromyosin (LMM), and the rest of the molecules, containing the two heads and the 43 nm-long rod, known as heavy meromyosin (HMM). HMM can be further split into two separate heads (subfragment-1 or S-1) and the rod (subfragment-2, S-2). In myosin (thick) filaments, LMM aggregates to form the filament backbone, which is polarized in opposite directions on either side of the filament central point, while myosin heads (S-1) extend laterally from the filament backbone with an axial interval of 14.3 nm, except for a region without myosin heads, i.e., the bare zone at the center of myosin filament (Fig. 4B). The S-2 rod is believed to serve as a hinge between myosin heads and the myosin filament backbone, thus enabling myosin heads to swing away from the myosin filament.

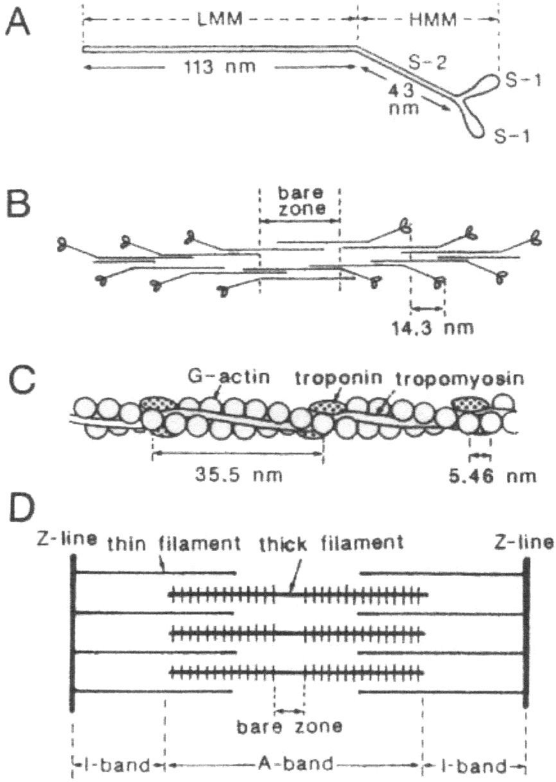

Figure 4 Structure of myosin and actin molecules and their formation of myosin and actin filaments. (A) Diagram of myosin molecule. (B) Arrangement of myosin molecules in myosin filament. (C) Structure of actin filament. (D) Arrangement of actin and myosin filaments in a sarcomere. Note that half-sarcomere is the structural and the functional unit of striated muscle. From Sugi (1992).

On the other hand, actin (thin) filaments primarily consist of two helical strands of globular actin monomers (G-actin, MW 41,700), which are wound around each other with a pitch of 35.5 nm. The axial separation of actin monomers in the actin filament is 5.46 nm (Fig. 4C). In the vertebrate skeletal muscle, actin filaments contain tropomyosin and troponin, whose function will not be discussed in this book. Owing to the symmetrical sarcomere structure, the direction of filament sliding is reversed across the bare zone (Fig. 4D). Since the establishment of the sliding filament mechanism in muscle contraction, the remaining central problem in understanding the molecular mechanism of muscle contraction is" What makes the filaments slide past each other?"

As already mentioned, H. E. Huxley got the idea that the cross connections, which he observed in his early electron microscopic observation of actin and myosin filaments (Fig. 2), and later proved to be myosin heads extending from myosin filaments, play an essential role in producing the filament sliding. In 1969, he put forward a hypothesis shown in Fig. 5.

Owing to the difference in their axial periodicity, the cyclic attachment–detachment between actin and myosin filaments takes place asynchronously. In the upper diagram, a myosin head (located on the left) first attaches to a site on an actin filament. Then, the attached head changes its configuration to produce a unitary sliding between the filaments (middle diagram). The direction of filament sliding is such that the sarcomere shortens. After producing the unitary filament sliding, the myosin head detaches from the site on actin filament (lower diagram). It is assumed that hydrolysis of one ATP molecule is coupled with each attachment–detachment cycle. It should be noted that after producing the filament sliding, myosin heads are assumed to return to their original configuration; the first myosin head movement producing the filament sliding is called the power stroke, while the subsequent myosin head movement to restore its original configuration is called the recovery stroke. In order to repeat power and recovery strokes, the two strokes should be the same in amplitude, and opposite in direction. This point will be discussed later.

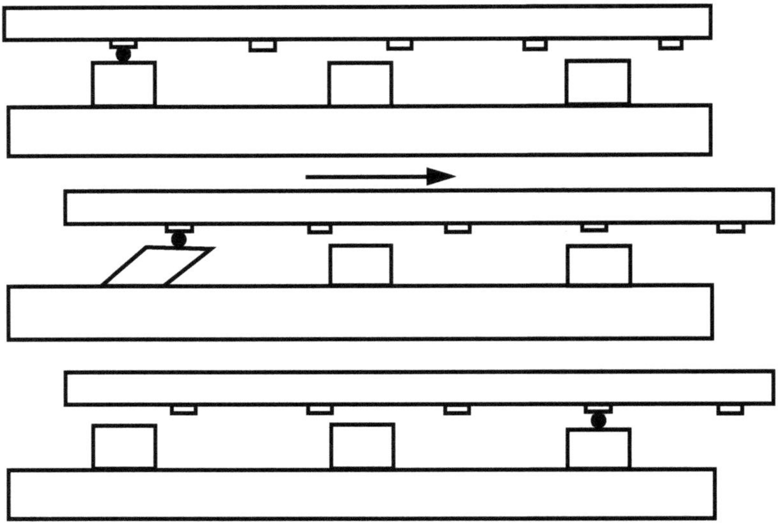

Figure 5 Proposed attachment–detachment cycle between myosin head extending from myosin filament and corresponding myosin-binding site on actin filament. From Huxley (1969).

Although more than 60 years have passed since the monumental discovery of the sliding filament mechanism in muscle contraction, the movement of myosin heads producing muscle contraction still remains a matter for debate and speculation.

1.4 A. F. Huxley's Contraction Model

Based on the findings described above, A. F. Huxley constructed a contraction model, which could explain most of the physiological, thermal, and biochemical results obtained on the skeletal muscle and muscle fibers, and has been central in the research field of muscle contraction up to the present time (A. F. Huxley, 1957).

A. F. Huxley's contraction model explains basic characteristics of contracting vertebrate skeletal muscle, which are summarized in Fig. 6. A muscle is regarded to consist of a contractile component (CC) and series elastic component (SEC) connected in series (Fig. 6B). When a muscle is fully activated to contract isometrically with both ends fixed in position, CC shortens internally to stretch SEC by ΔL_{max}, while generating isometric force P_0. If the isometrically contracting muscle is rapidly released by $\Delta L < \Delta L_{max}$, the isometric force drops from P_0 to P and then starts developing toward P_0 (Fig. 6A). If the force in contracting muscle is quickly reduced from P_0 to P, the muscle first shows rapid length change ΔL and then starts shortening with a constant velocity (Fig. 6D). The constant velocity shortening under a constant load (=force) P, i.e., shortening without acceleration, is a most prominent characteristic of the contracting muscle. Hill has shown that the relation between force P (=load) and the shortening velocity V can be expressed as a part of the rectangular hyperbola (Fig. 6E) (Hill, 1938). Meanwhile, the force-extension curve of SEC is parabolic in shape, with ΔL_{max} about 1% of muscle slack length L_0 (Fig. 6C).

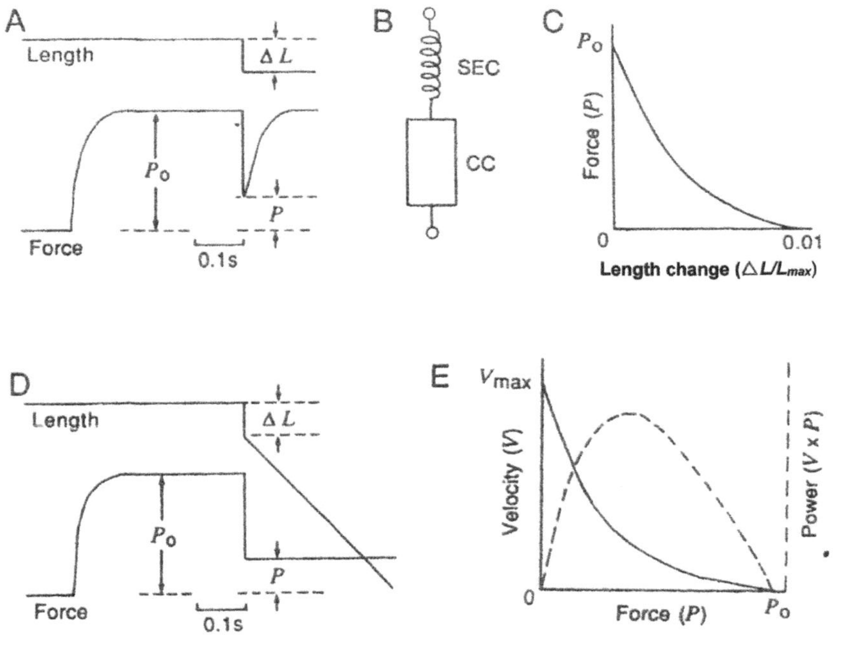

Figure 6 Basic characteristics of contracting skeletal muscle. From Sugi (1992).

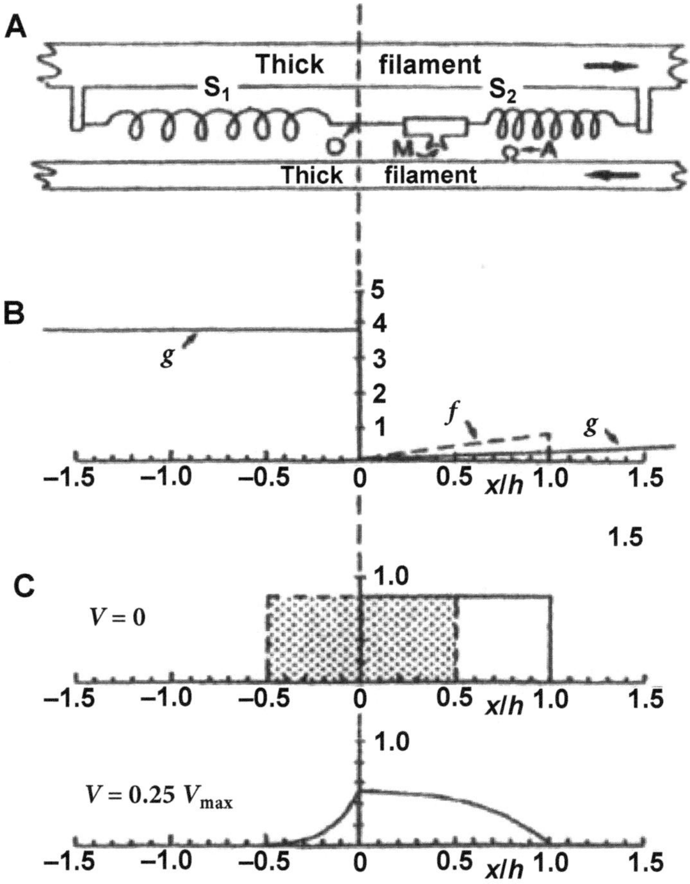

Figure 7 A. F. Huxley's contraction model. For further explanation, see text.

Figure 7 illustrates the outline of A. F. Huxley's contraction model. A myosin head is connected to a thick (myosin) filament backbone via springs S_1 and S_2 and fluctuates around the equilibrium position 0 due to thermal motion (Fig. 7A). When M attaches to A in the contracting muscle, it generates positive force for filament sliding at the right side of 0, and negative forces at the left side of 0. Both positive and negative forces are proportional to distance x from 0. The rate constants for both making and breaking the A–M link, f and g, respectively, are functions of x (Fig. 7B). The value and distribution of these rate constants are chosen to fit Hill's experimental results. In the isomeric condition, in which a muscle generates isometric force P_0 without shortening, A–M links are formed in the region where $f > g$, i.e., $0 \leq x \leq h$ (Fig. 7C). If the contracting muscle is rapidly released by $h/2$, the A–M link distribution shifts to the left, so that the positive forces generated by the A–M link in the positive x region become equal to the negative force generated by the A–M link in the negative x region (shaded area in the upper diagram in Fig. 7C). The value $h/2$ corresponds to ΔL_{max}, i.e., the amount of rapid release to drop isometric force P_0 to zero.

During steady shortening under load (=force) P with a constant velocity, a number of M pass through A from right to left along the x-axis. In the positive x region, the formation of the A–M link occurs at $0 \leq x \leq h$, and then A–M links decrease to zero in the negative x region, where g is high and $f = 0$. The A–M link distribution during steady muscle shortening under a load (=force) of 0.25 P_0 is shown in the lower diagram in Fig 7C. The difference between the positive force in the positive x region and negative force in the negative x region is equal to P, to be measured externally.

Chapter 2

Dogmas of Huxley–Simmons Model Not Supported Experimentally

The history of natural science tells us that monumental breakthroughs are sometimes achieved by pioneers to open new research field. As the result, many investigators start studying their experimental objects, found by pioneers. Because a pioneering work is not necessarily perfect, it includes dogmas, which are to be modified or altered in future. Nevertheless, a majority of investigators, gathered in the new research field, are not aware that the dogmas are wrong and continue their research work along the way shown by wrong dogmas. As the result, the progress in this particular research field is retarded. This period is called "normal science." In this chapter, I will describe some examples of how wrong dogmas were broken and how other wrong dogmas are still believed to be correct from my standpoint.

2.1 Huxley–Simmons Contraction Model: Dogmas about Myofilament Rigidity and Myosin Head Tilting

In 1971, A. F. Huxley and R. M. Simmons presented a new contraction model, in which they made detailed structural

Mysteries in Muscle Contraction: Evidence against Current Dogmas
Haruo Sugi

ISBN 978-981-4774-48-2 (Hardcover), 978-1-315-10191-0 (eBook)
www.panstanford.com

considerations about myosin head power stroke (A. F. Huxley and Simmons, 1971, 1972). They recorded force response of a single skeletal muscle fiber, using the segmental clamp technique, with which the length of a short muscle fiber segment was controlled with a feedback circuit to avoid complications arising from non-uniform length changes along the fiber length.

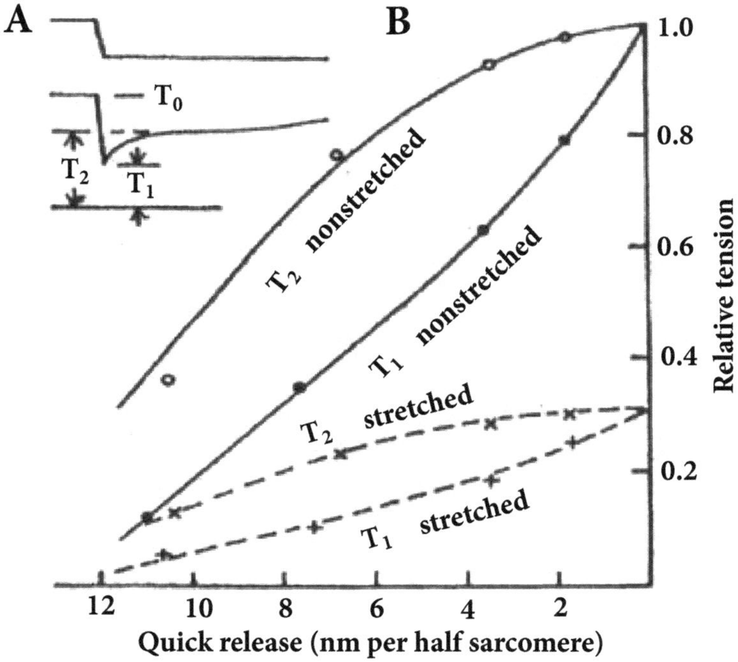

Figure 8 (A) Force (tension) response (lower trace) of a single muscle fiber to a quick release (upper trace). (B) Dependence of T_1 and T_2 on the amount of quick release in non-stretched and stretched fibers. From Huxley and Simmons (1972).

Figure 8 illustrates their experimental results, which led them to propose a new contraction model. They applied quick changes in the fiber length (quick release) on isometrically contracting muscle fibers generating steady isometric force T_0. In response to a quick release, the fiber exhibited initial elastic force drop from T_0 to T_1, coincident with the applied release, which was followed by quick force recovery from T_1 to T_2. Then the force continued to rise towards T_0. They thought that the quick force recovery reflects redistribution of myosin heads, attaching to the actin filament at different angles. To prove the validity of their idea, they applied quick releases to the fiber at non-stretched and stretched fiber lengths to alter the number of myosin heads involved in contraction by changing the amount of overlap between actin and myosin filaments. Despite the marked reduction of T_0 in stretched fibers, the dependence of T_1 and T_2 on the amount of quick release appeared to be identical; both T_1 and T_2 curves were scaled down according to the value of T_0. They concluded that force generation in the muscle results from tilting (or change in angle) of myosin heads attached to the actin filament as illustrated in Fig. 9.

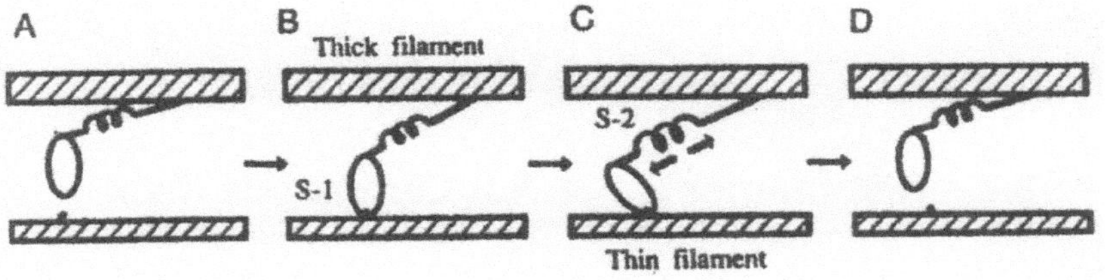

Figure 9 Diagram showing attachment–detachment cycle between myosin head (S-1) and thin (actin) filament. (A) Myosin head detached from actin filament. (B) Attachment of myosin head to actin filament at angle of 90°. (C) Change in attachment angle of myosin head from 90° to 45°, by stretching myosin subfragment-2 (S-2) connecting myosin head to thick (myosin) filament backbone. (D) Detachment of myosin head from actin filament. The force generated by stretched S-2 appears externally as force in muscle. From Sugi (2006).

In isometrically contracting muscle, myosin heads first attach to actin filaments at an attachment angle of 90° (A to B) and then change their attachment angle from 90° to 45° (B to C) by stretching the link, i.e., myosin subfragment-2, connecting myosin heads to the myosin filament backbone. The force generated by the stretched S-2 appears externally as isometric force in contracting muscle. After the tilting, i.e., change in attachment angle, myosin heads detach from actin filaments (C to D). The attachment–detachment cycle continues in contracting muscle to generate steady isometric force.

In the above contraction model, both actin and myosin filaments are assumed to be completely rigid, and therefore muscle fiber elasticity is assumed to entirely reside in the S-2 region of myosin. The validity of these assumptions depends on whether the T_1 versus ΔL (amount of quick release) curve actually intersects ΔL axis at ΔL_{max} or not. The electronic length-clamp device works well if the length versus force relation is steep, but does not work well if this condition is not fulfilled. For this reason, the intersection of T_1 versus ΔL curve with the ΔL axis was not clear.

Despite the above incomplete experimental basis, it has been believed for many years that (1) both actin and myosin filaments are completely rigid and (2) muscle fiber elasticity resides in myosin heads. As a consequence of these dogmas, it has been believed that the amplitude of myosin head power stroke is about 1% of the muscle fiber length or about 10 nm per half sarcomere.

2.2 Myofilaments Are Not Completely Rigid

2.2.1 Electron Microscopic Studies on the Extensibility of Myofilaments

Soon after the proposal of contraction model by A. F. Huxley and Simmons in 1971, I wondered about the complete rigidity of actin and myosin filaments in the A. F. Huxley–Simmons contraction model, and performed experiments with Dr. Suzuki, in which changes in sarcomere structures in rigor muscle fibers were observed electron microscopically before and after stretch of rigor muscle fibers.

In the absence of ATP, myosin heads and actin filaments form tight rigor linkages. Consequently, when the muscle fibers in the rigor state are stretched appropriately, myofilaments located out of the rigor linkage region would be extended to show their finite elasticity (Suzuki and Sugi, 1983). In intact muscle fibers, it is difficult to change chemical and ionic conditions in the interior of the fiber because of the surface membrane. This difficulty is overcome by using skinned muscle fibers, in which the surface membrane is removed by mechanical or chemical means (Brenner, 1998). We used skinned fibers prepared by treatment with glycerol, so that rigor actin-myosin linkages were readily produced by immersing skinned fibers in ATP-free solution.

Electron micrographs of myofibrils in rigor fibers, fixed before and after 10% stretch, are shown in Fig. 10. It can be seen that the width of the I-band and the bare region is increased markedly, while the width of the region where actin and myosin filaments overlap and form tight rigor linkages does not change appreciably. Based on the force in stretched rigor fibers and degree of actin and myosin filament extension, we calculated approximate values of myofilament extensibility at the bare region, and actin filament extensibility at the I-band; in terms of stiffness, myosin filament stiffness per unit length (at the bare region) was ~3.6×10^4 pN, while actin filament stiffness per unit length was ~2.4×10^4 pN (Suzuki and Sugi, 1983). The actin filament stiffness was nearly equal to the value (1.8×10^4 pN) assumed by White and Thorson (1973) taking the limit of accuracy of X-ray diffraction measurement of actin filament length into consideration. The total contribution of myofilament elasticity to the SEC was ~0.4% of Lo, amounting ~40% of ΔL_{max}, indicating that myofilament stiffness contributes significantly to the value of SEC.

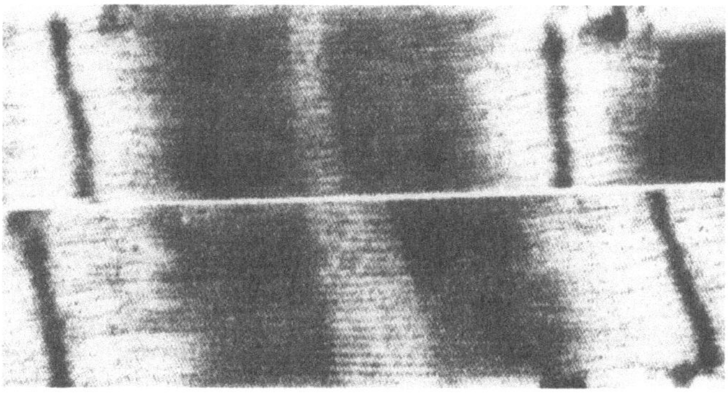

Figure 10 Electron micrographs of sarcomeres before (upper panel) and after 10% stretch (lower panel) of rigor muscle fibers. Note marked elongation of bare region and I-band, while the width of myofilament overlap region does not change appreciably. From Suzuki and Sugi (1983).

Meanwhile, Ford et al. (1981) estimated actin filament stiffness to be ~3×10^5 pN, a value more than 10 times larger than the values obtained by us and White and Thorson. According to them, the contribution of myofilament stiffness amounted to only ~10% of ΔL_{max}.

2.2.2 Electron Microscopic Studies on the Extensibility of Actin Filament Extensibility Using Freeze-Fracture Method

I made a more accurate electron microscopic study with Drs. Suzuki and Oshimi (Suzuki et al., 1993) on the extensibility of actin filaments, by examining freeze-etch replicas of quickly frozen rigor muscle fibers. Figure 11 illustrates the experimental arrangement. The middle portion of a skinned muscle fiber, with its surface membrane removed, was placed between two hat-shaped gold plates P1 and P2 in such a way that the fiber ran between the two plates without being compressed or deformed. Immediately before quick freezing, the solution in the experimental chamber was drained, and then the fiber was quickly frozen by jetting liquid propane (−190°C) from two nozzles N1 and N2 (Suzuki and Pollack, 1986). The methods to prepare freeze-etch replicas of the fiber have been described elsewhere (Suzuki et al., 1993).

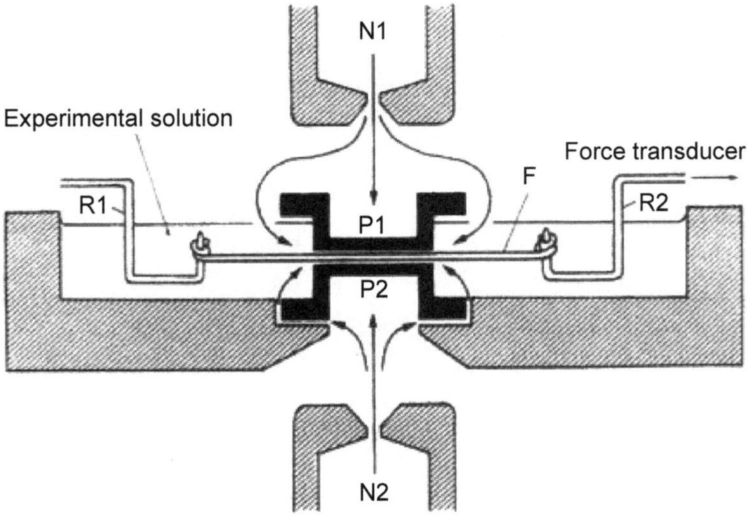

Figure 11 Experimental setup to quickly freeze single skinned muscle fibers. One end of the fiber was connected to a force transducer, while the other end was connected to a servo motor to apply length changes to the fiber. From Suzuki et al. (1993).

We examined the variation of axial spacing of actin monomers in actin filaments, in the myofilament nonoverlap region, i.e., the I-band, among three different rigor states; (1) during generation of isometric rigor force (~0.2 Po), (2) after stretch to increase rigor force to ~0.6 Po, and (3) after release to drop rigor force to zero. As shown in Fig. 12, the axial spacing of actin monomer increased significantly ($P < 0.01$) with increasing rigor force immediately before quick freezing. Based on some assumptions, the elasticity of actin monomer was estimated to be ~1.8×10^4 pN per unit length, a value not much different from that of actin filament elasticity (~2.4×10^4 pN per unit length) obtained by Suzuki and Sugi (1983).

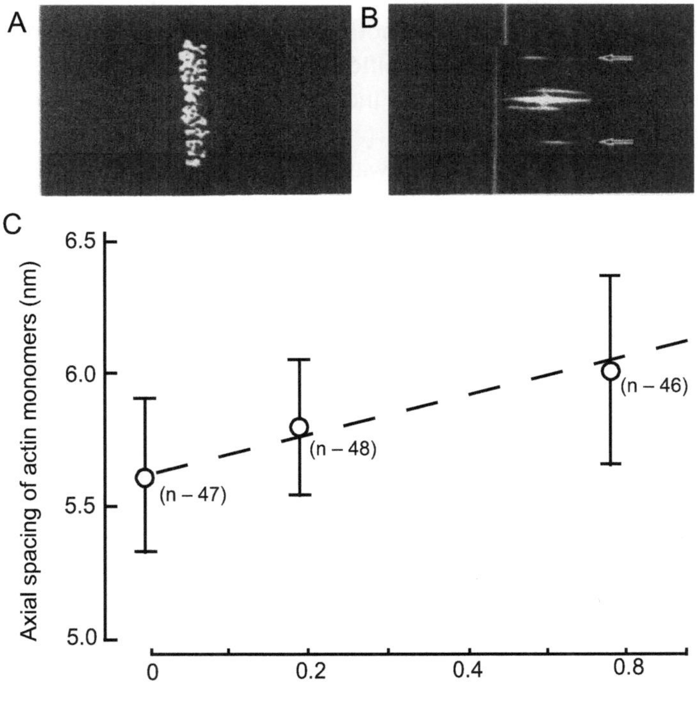

Figure 12 (A) An example of actin filament segment replica image obtained from the I-band region. (B) Optical diffraction pattern from the image shown in A. Layer lines based on axial spacing of actin monomer are indicated by arrows. (C) Relation between rigor force in the fiber and axial spacing of actin monomer (mean ± SD). From Suzuki et al. (1993).

We prepared a paper reporting the above results and submitted it to a mainstream journal, but it was rejected by an editor citing the reason that "*Myofilaments are already shown to be completely rigid according to Huxley contraction model. You should have made serious mistakes in conducting experiments*". I realized the strong influence of the dogma of myofilament rigidity even almost 25 years after the publication of the Huxley contraction model in 1957.

One year after our bitter experience, however, the dogma was denied definitely. Using the intense X-rays from synchrotron radiation, H. E. Huxley et al. (1994) and Wakabayashi et al. (1994) performed X-ray diffraction studies on isometrically contracting intact muscle with a high spatial resolution and showed convincingly that both actin and myosin filaments have finite elasticity, which significantly contributes to the elasticity of the contracting muscle, indicating that the amplitude of the myosin head power stroke (about 1% of fiber length, or about 10 nm per half-sarcomere) was an overestimate. The myofilament elasticity estimated by them was not much different from that obtained by us in 1983 and 1993.

2.3 Myosin Head Tilting Was Not Supported Experimentally

Meanwhile, a number of experiments were made to detect myosin head tilting, predicted by the A. F. Huxley–Simmons contraction model. The equatorial X-ray diffraction of muscle arises from a hexagonal myofilament-lattice, and the intensity of equatorial reflections is sensitive to the mass distribution within its unit cell. Since muscle force generation is expected to result from myosin head tilting (Fig. 8), quick length changes applied to the rigor muscle, in which all myosin heads attach to actin filaments, are expected to change the angular distribution of attached myosin heads to affect equatorial reflection intensities. The stretch of rigor muscle fibers, however, produced no marked changes in equatorial reflection intensities (Naylar and Podolsky, 1981; Tanaka, Sugi, and Hashizume, 1984), indicating that the angular distribution may not be readily altered by the stretch. Using the technique of time-resolved X-ray diffraction, Amemiya et al. (1980) applied a quick release (1% of muscle length) to the isometrically contracting muscle to drop isometric force to zero but could not detect any appreciable changes in equatorial reflection intensities. Huxley et al. (1983) also obtained similar results. Using chemical probes attached to myosin heads in skinned muscle fibers, Yanagida (1984, 1985), Cooke (1981), and Cooke, Crowder and Thomas (1982) also failed to detect appreciable changes in myosin head orientation in skinned rigor muscle fibers. Thus, the idea of myosin head tilting as the cause of myofilament sliding (Fig. 8) was not supported by a number of experiments.

2.4 Quick Force Recovery Following Quick Release Is a Complex Viscoelastic Phenomenon

As described above, the quick force recovery, which was regarded as being due to redistribution in the angle of attachment of myosin heads caused by quick release, does not now have a concrete basis, because this idea was coupled with the myosin head tilting mechanism, which is not supported experimentally. At the end of this chapter, I will briefly describe the results of experiments performed by Sugi and Kobayashi (1983), since I believe that this work is valuable for future investigators as a basis to challenge the remaining mysteries in muscle contraction mechanisms.

As soon as A. F. Huxley and Simmons presented their new contraction model in 1971, I felt skeptical about the uniformity of mechanical events following quick release if the propagation velocity of mechanical impulse along the fiber length was taken into consideration. We constructed an experimental setup shown in Fig. 13.

This setup can record small sarcomere length changes by splitting the first-order diffraction beam of He-Ne laser light originating from sarcomere spacing at various regions of the muscle fiber generating a steady isometric force. Quick releases are applied at one fiber end connected to the servo motor, while the resulting force changes in the fiber are recorded at the other fiber end fixed in position (Sugi and Kobayashi, 1983).

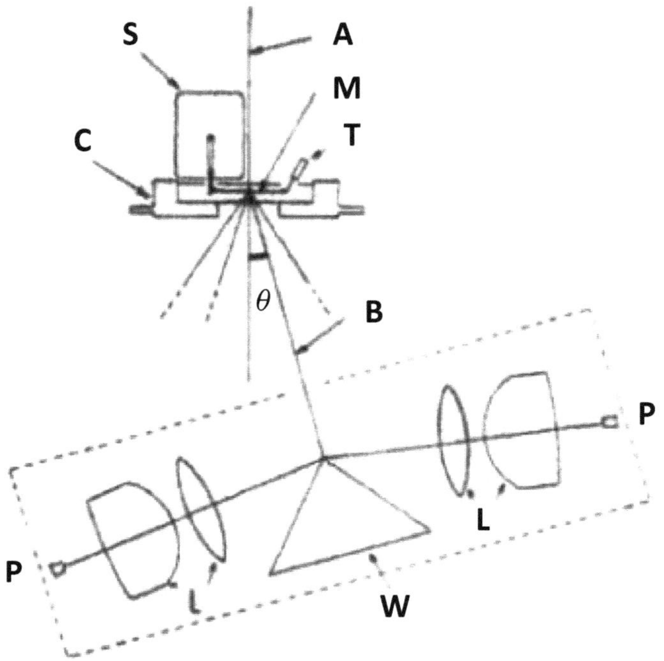

Figure 13 Experimental setup. A muscle fiber (M) is mounted horizontally in experimental chamber (C) between servomotor (S) and force transducer (T) and stimulated to contract isometrically. Small sarcomere length changes are recorded as a small shift of the first-order diffraction line of HE/Ne laser beam (A) by splitting the diffraction beam (B) with a wedge-shaped mirror (W) into two parts, each of which is focused on a photodiode (P) through lens system (L). For small sarcomere length changes, voltage difference between two diode signals is a linear function of angle (θ) between the incident and the first-order diffraction beams. From Sugi and Kobayashi (1983).

A typical result is presented in Fig. 14. The upper record shows the time course of quick release applied at the region near the moving fiber end, connected to the servo motor (recorded with a displacement transducer incorporated in the servo motor), the middle record shows the time course of sarcomere length changes recorded with the wedge-shaped mirror system in the region near the fixed fiber end, connected to the force transducer, and the lower record shows the time course of force changes recorded with the force transducer at the fixed fiber end. It can be seen that (1) there is a definite difference in the time of onset of length changes between the region near the moving fiber end and the region near the fixed fiber end, due to the propagation of mechanical impulse along the fiber (~180 m/s; Schoenberg, Wells and Podolsky, 1974); (2) sarcomere length shortening near the fixed fiber end outlasts the fiber shortening near the servomotor; and (3) force changes, recorded with the force transducer at the fixed fiber end, indicate that the quick force recovery starts, while sarcomere shortening near the fixed fiber end is still in progress.

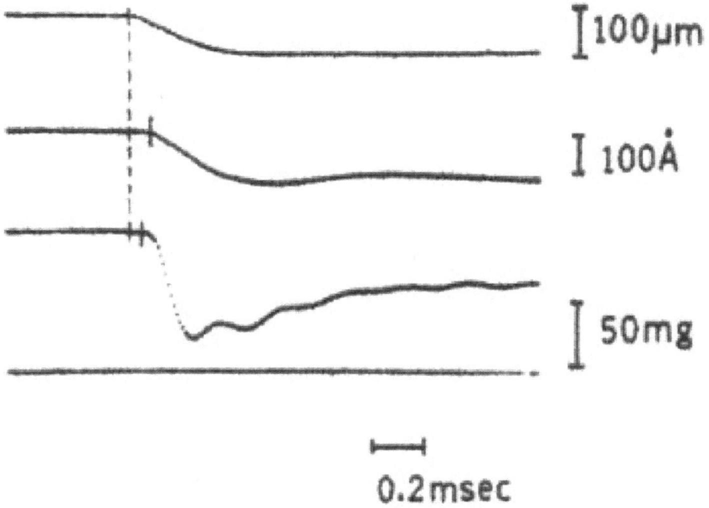

Figure 14 Simultaneous recordings of fiber length changes recorded at the moving fiber end (upper trace), sarcomere length changes recorded near the fixed fiber end (middle trace), and force changes recorded at the fixed fiber end (lower trace). Note that quick force recovery recorded by the force transducer at the fixed fiber end occurs while sarcomere shortening near the fixed fiber end is still in progress. From Sugi and Kobayashi (1983).

These results indicate that the quick force recovery following quick release is a complicated viscoelastic phenomenon associated with non-uniform length changes of fiber segments. We have constructed a multi-segmental model consisting of 20 of sarcomere elements connected in series. Based on our electron microscopic observation of stretched rigor muscle fibers (Fig. 10), each sarcomere element is composed of series elastic elements (SEC), corresponding to actin filaments in the I-band and myosin filaments in the bare region, and contractile, viscous, and elastic elements connected in parallel, corresponding myofilament overlap region in the A-band (Fig. 15). By giving a rapid displacement to one end of the multi-segmental model, we could qualitatively simulate the results shown in Fig. 16 (Nishiyama, 1974); when a rapid length change is applied at the moving end of the model, the resulting sarcomere shortening outlasts the quick force recovery.

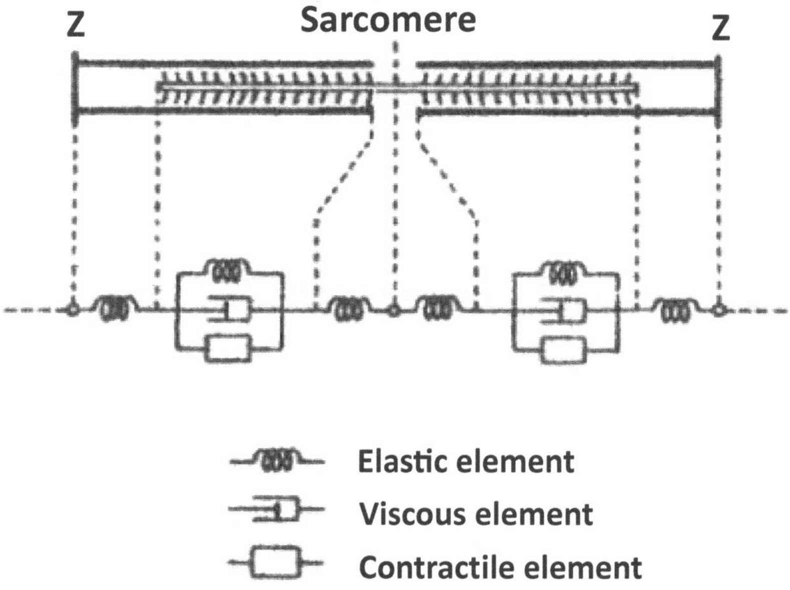

Figure 15 Elements of multi-segmental model constructed on the basis of electron microscopic observation shown in Fig. 10. From Nishiyama and Sugi (1974).

Our work described above indicates that, to obtain useful information about the mechanism of muscle contraction, multi-segmental nature of muscle fibers should be taken into consideration; the inherent problem in muscle mechanics is that it is possible to clamp the segmental length of muscle fiber, but it is not possible to clamp the "segmental force" within a short time period, when the propagation of mechanical impulse along the fiber length is in progress.

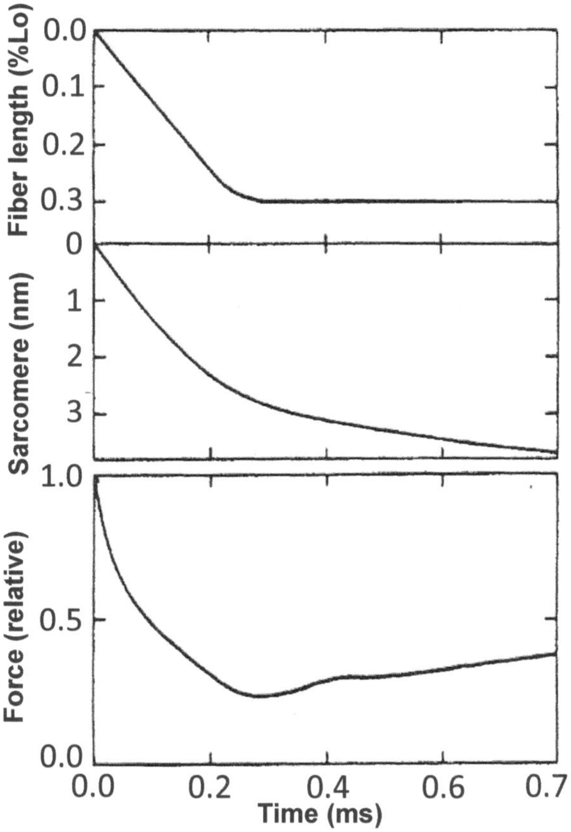

Figure 16 Simulation of sarcomere length and force (tension) changes at the fixed fiber end in response to a rapid length change applied to the moving fiber end. Upper trace, time course of rapid length change at the moving fiber end. Middle trace, time course of sarcomere length change at the fixed fiber end. Lower trace, time course of force changes at the fixed fiber end. From Nishiyama and Sugi (1974).

Chapter 3

New Methods and New Dogmas and Their Shortcomings

In this chapter, I will explain how current dogmas on the mechanism of muscle contraction at the molecular level have been constructed on a variety of experimental results, resulting from the development of techniques in studying muscle contraction. All the current dogmas are concerned with structural changes of myosin molecules, especially nucleotide-dependent structural changes of myosin head crystals. From the standpoint of muscle physiology, although these crystallographic studies are in themselves very interesting, the results obtained appear not to give direct answers to the question: "What is actually taking place in the 3D myofilament-lattice in muscle?" To begin with, I will explain the Lymn–Taylor scheme on the kinetics of actomyosin ATPase reaction, and then describe the crystallographic studies on myosin head crystal as well as cryo-electron microscopy of the acto-S1 complex, on which current dogmas are constructed.

3.1 The Lymn–Taylor Scheme of Actomyosin ATPase Reaction

The molecular mechanism of muscle contraction can be approached biochemically by studying the ATPase reactions of

Mysteries in Muscle Contraction: Evidence against Current Dogmas
Haruo Sugi

ISBN 978-981-4774-48-2 (Hardcover), 978-1-315-10191-0 (eBook)
www.panstanford.com

actin and myosin mixture, i.e., actomyosin complex, in solution. In order to relate the enzyme kinetics of actomyosin in solution to the characteristics of muscle contraction as described in the preceding chapters, however, the following problems and limitations should be kept in mind.

(1) The function of the 3D myofilament lattice to convert the energy derived from ATP hydrolysis into mechanical work is lost in the actomyosin solution, so that all the free energy of chemical reactions is dissipated as heat and entropy.
(2) No theory has hitherto been developed to deal with the probability of reactions between myosin heads and actin filaments taking place in the 3D myofilament lattice in the muscle.
(3) Both the ionic strength and effective contractile protein concentrations are much higher in the muscle than in solutions, in which enzyme kinetics can be studied.

Despite these limitations, it is possible to some extent to relate biochemical results to the mechanism of muscle contraction.

As already mentioned, myosin heads have the ATP-binding site and the ATPase activity. Their normal substrate in solution is Mg-ATP. The kinetics of myosin ATPase is studied by examining the time course of ATP hydrolysis by myosin, HMM, or S-1 extracted from rabbit skeletal muscle. For the sake of convenience, the character M will hereafter be used to denote myosin head (S-1), in which the ATPase activity and the actin-binding site are localized, irrespective of whether myosin, HMM, or S-1 is used.

Using the quenched flow apparatus (Fig. 17a), Tonomura et al. (1962) found that on mixing ATP with M, there was a very rapid liberation of inorganic phosphate (Pi), i.e., phosphate burst, which was followed by a much slower Pi liberation. Using the stopped flow apparatus (Fig. 17b), Lymn and Taylor (1970) explained the above phenomenon as being due to rapid ATP hydrolysis to produce ternary complex, M-ADP-Pi, which was indistinguishable from Pi in the quenched flow apparatus.

Although the formation of M-ADP-Pi on mixing ATP and M is rapid, the subsequent steady release of ADP and Pi from M is slow (~0.01/s; Bagshaw, 1994). If, however, ATP is mixed with acto-S-1 or acto-HMM, i.e., mixture of S-1 or HMM and actin, the amount of initial Pi burst is similar to that when only M is mixed with ATP, while subsequent steady-state ATPase activity increases more than 200 times (Eisenberg and Moos, 1968, 1970). This indicates that the rate-limiting step, i.e., release of ADP and Pi from M, is accelerated by actin.

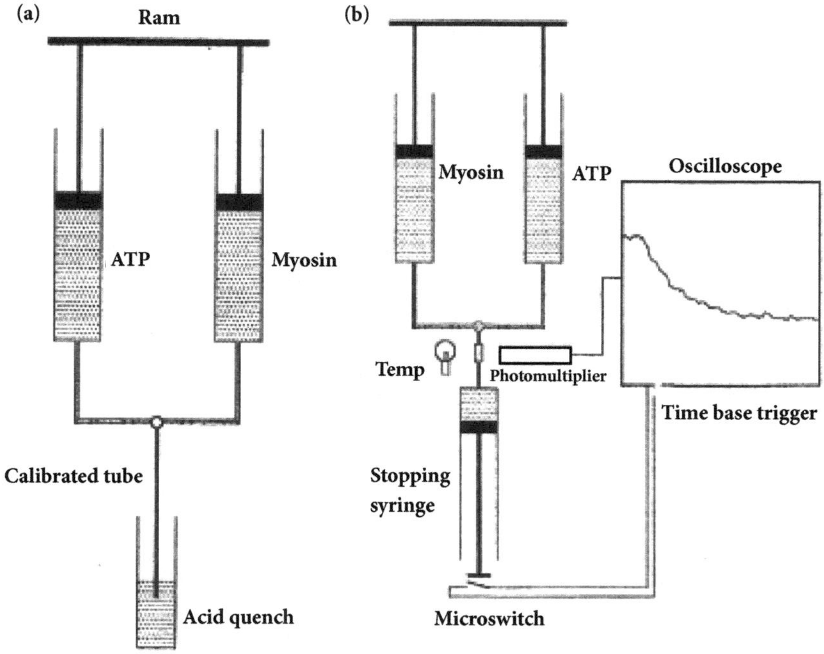

Figure 17 (a) Quenched flow apparatus, in which ATP and myosin are rapidly mixed into quenching agent (e.g., acid), and reaction products are analyses. (b) Stopped flow apparatus, in which ATP and myosin are rapidly mixed into a small cuvette, where formation of reaction products are recorded as optical signals. From Bagshaw (1994).

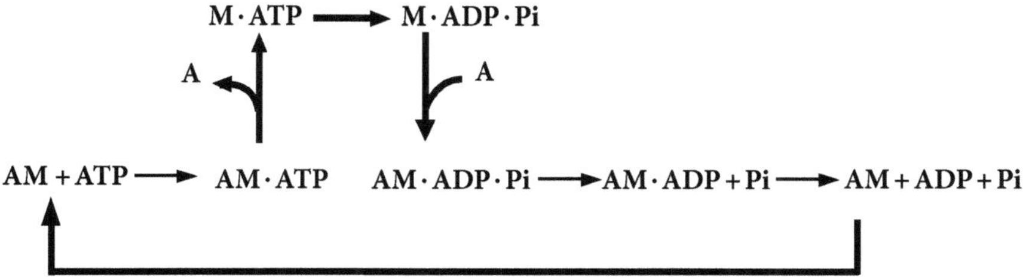

Figure 18 The Lymn–Taylor scheme of cyclic reaction steps. In each cycle, one ATP molecule is hydrolyzed into ADP and Pi.

Taking these and other experimental results into consideration, the most probable sequence of the reaction taking place in contracting muscle (the Lymn–Taylor scheme) is shown in Fig. 18. The reaction cycle includes the attachment of actin (A) to, and its detachment from, the actin-binding site on myosin head (M), thus providing a simple, clear-cut correspondence to the contraction models described in the previous chapters (Figs. 5 and 9). That is why the Lymn–Taylor scheme has been cited by muscle physiologists in many papers on the mechanism of muscle contraction.

In relaxed muscle, most myosin heads (M) are believed to be in the state M-ADP-Pi, having the average lifetime >10 s; the attachment of M to A is inhibited by the action of regulatory proteins, tropomyosin and troponin (Fig. 4C). In contracting muscle, in which the inhibitory action of regulatory proteins is removed by Ca ions, M in the form of M-ADP-Pi attaches to A to form A-M-ADP-Pi, which releases Pi and ADP in two steps, i.e., (1) A-M-ADP-Pi → A-M-ADP + Pi and (2) A-M-ADP → A-M + ADP. During the release of Pi and ADP, M firmly attaches to A to perform the power stroke producing unitary myofilament sliding (see Fig. 5) when contracting muscle shortens, or generates isometric force.

After the completion of the power stroke, M remains firmly attached to A to form complex A-M, until the next ATP comes to bind with M. On binding with ATP, M detaches from A as, A-M + ATP → A-M-ATP → M-ATP + A, and performs the recovery stroke to restore its original configuration, associated with the reaction M-ATP → M-ADP-Pi. After the completion of the recovery stroke, M again attaches to A in the form of A-M-ADP-Pi. Thus, the attachment–detachment cycle between M and A is repeated.

In order that M repeats attachment–detachment with A, coupled with conformational changes in M, the power and the recovery strokes should be the same in amplitude and opposite in direction. This is proved to actually be the case by our electron microscopic recording of ATP-induced myosin head movement, which will be described in this book later. In the Lymn–Taylor scheme, the A-M binding in A-M-ADP-Pi is believed to be weak, while the A-M binding in A-M-ADP and A-M is strong. Consequently, the myosin head power stroke should be associated with the transition from weak to strong binding states between A and M.

3.2 Electron Microscopic Studies on the Static Structure of Rigor Actin–Myosin Complex

It is well known that in the absence of ATP, actin (A) and myosin head (M) form a high-affinity complex A-M, corresponding to the rigor state of muscle. To obtain information about actin–myosin interaction at the molecular level, the structure of rigor A-M complex has been studied intensively using an electron microscope.

Figure 19 shows the 3D reconstruction of unstained, quickly frozen acto-S1 complex, in which a number of myosin heads (S-1) are attached around an actin filament. The static structure of the acto-S-1 complex gives, however, no direct information about the dynamic interaction of myosin heads and actin filaments taking place in contracting muscle.

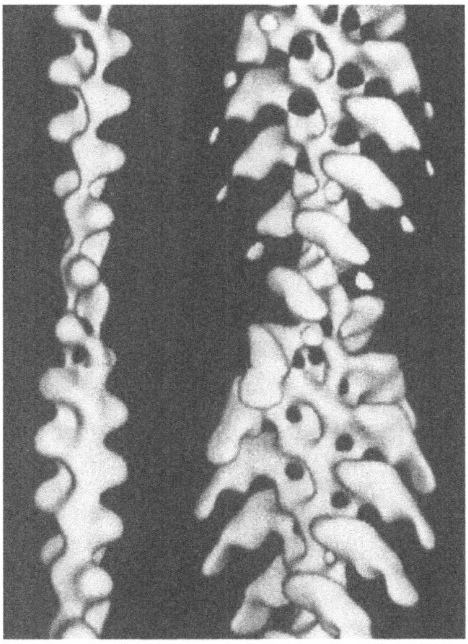

Figure 19 3D reconstruction of cryo-electron micrograph of actin filament decorated with myosin heads (acto-S1 complex). From Milligan et al. (1990).

We have, however, physiological evidence that myosin heads and actin filaments never pass through the static structure shown in Fig. 19 during their dynamic interaction in the contracting muscle.

As illustrated in the simplified diagram of the primary structures of myosin head-actin molecule complex (Fig. 20), major myosin-binding sites on actin are localized near its N terminus, while the corresponding actin-binding sites on the myosin head are localized at both sides of the junction between 50 K and 20 K fragments of the myosin heavy chain.

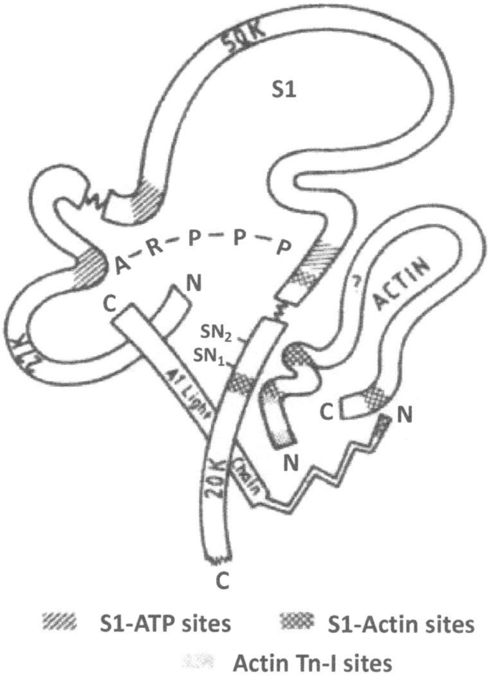

Figure 20 Simplified diagram of primary structures of myosin head-actin molecule complex, showing their likely points of interaction on their amino acid sequence. From Levine et al. (1990).

It should be noted, however, that both the structure of acto-S-1 complex (Fig. 19) and the actin-binding sites of the myosin head (Fig. 20) are obtained from protein samples extracted from muscle, and they are therefore not necessarily related to structures and actual manner of binding of myosin heads and actin filaments in the 3D myofilament-lattice in muscle.

Nevertheless, a dogma is constructed that in contracting muscle, (1) the state A-M in the Lymn–Taylor scheme corresponds, or is analogous to, the acto-S-1 complex (Fig. 19), and (2) myosin heads pass through the A-M state, in which myosin heads bind tightly with actin via their binding sites (Fig. 20) to form the rigor complex (Fig. 19), which is determined by cryo-electron microscopic studies using extracted protein samples. Again, it appears likely that such a dogma, based on the static acto-S-1 complex structure, cannot be applied to the 3D myofilament-lattice in the contracting muscle.

3.3 Development of in vitro Motility Assay Systems

In order to eliminate the gap between muscle physiology and muscle biochemistry, a variety of experimental systems have been developed, in which ATP-dependent sliding between actin and myosin can be studied quantitatively. These experimental systems are called the in vitro motility assay systems. Here, I will discuss the results obtained from these systems, which are relevant to the amplitude of the myosin head power stroke.

3.3.1 Fluorescent Actin Filament versus Myosin System

This experimental system was developed by Kron and Spudich (1986) and consists of fluorescently labeled actin filaments (F-actin) and myosin molecules, HMM, or S-1 fixed randomly on a glass surface. With this system, it was possible to observe ATP-dependent movement of fluorescent actin filaments over myosin under a light microscope. This excited muscle investigators to start using this assay system. Efforts were made to prove the amplitude of myosin head power stroke of ~10 nm per ATP molecule in the Huxley–Simmons contraction model. Unfortunately, the values obtained were extremely divergent, ranging from 10 nm to >100 nm, mainly due to problems in estimating the number of myosin heads involved in actin filament sliding and in accurately measuring the ATPase activity within a small experimental system. In addition, H. E. Huxley (1990) pointed out that the mechanism of the zigzag movement of actin filaments over randomly fixed myosin heads is not clear at all.

3.3.2 Actin Cable versus Myosin System

This assay system was developed by Sheez and Spudich (1983) and Shimmen and Yano (1984), and employs well-organized actin filament arrays occurring in the giant internodal cell of green algae, and small beads coated with myosin. The advantage of this assay system is that myosin-coated beads move straight along actin cables over a considerable distance. Here, I will describe our experiments, in which we examined whether the basic characteristics of the contracting muscle are retained in a small number of myosin molecules, i.e., myosin heads, interacting with actin filaments.

As shown in Fig. 21A, a myosin-coated glass microneedle is made in contact with actin cables in a strip of giant algal cell. In the absence of ATP, myosin heads form tight rigor linkages with actin. On iontophoretic application of a constant amount of ATP, myosin heads start moving along actin cables for a distance until applied ATP molecules are partly used up by myosin heads and partly diffuse away from myosin heads (Fig. 21B). To facilitate complete exhaustion of applied ATP, the experimental solution contains hexokinase and D-glucose serving as the ATP scavenger.

The ATP-induced movement of myosin-coated needle was recorded with a photoelectric device using a wedge-shaped mirror, similar to that used by us for recording minute sarcomere length changes in single muscle fibers (Fig. 13). We calculated the amount of work done by ATP-induced sliding between myosin-coated needle and actin cables as a function of initial baseline force at which needle started moving in response to ATP application. The relation between the amount of work done and initial baseline force was bell-shaped (Fig. 21C), indicating that a small number of myosin heads at the needle tip change their amount of work by utilizing a constant ATP supply depending on the initial condition, i.e., the amount of baseline force exerted by the bent needle. The above bell-shaped work versus initial force relation reminds us of the bell-shaped power versus force (=load) relation shown in Fig. 6E (bell-shaped broken line), indicating that a small number of myosin heads (probably <10) interacting with actin retain the basic characteristics of contracting muscle. This point will be discussed again later.

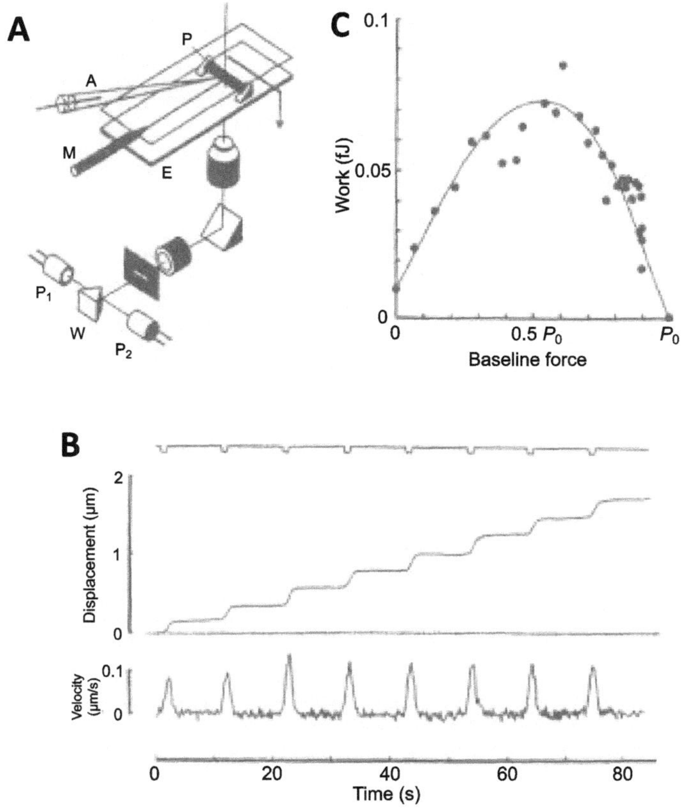

Figure 21 Measurement of work done by the ATP-induced sliding between a myosin-coated glass microneedle and actin cables. (A) Experimental arrangement. Myosin-coated needle (M) is in contact with actin cables on the algal cell strip (P). A constant amount of ATP is supplied iontophoretically to myosin from ATP-containing microelectrode (A). The needle tip image is sprit with wedge-shaped mirror (W) to be recorded with photodiodes (P_1 and P_2). (B) Simultaneous recordings of ATP current releasing constant amount of ATP from ATP electrode (upper trace), ATP-induced needle movement (middle trace), and velocity of needle movement (lower trace). (C) Relation between the amount of work done by the ATP-induced actin–myosin sliding and the initial baseline force at which needle starts moving. From Oiwa, Chaen, and Sugi (1991).

We also measured the velocity of sliding of myosin-coated polystyrene beads along actin cables under various external loads by using a centrifuge microscope. Figure 22 shows the experimental arrangement and the results obtained.

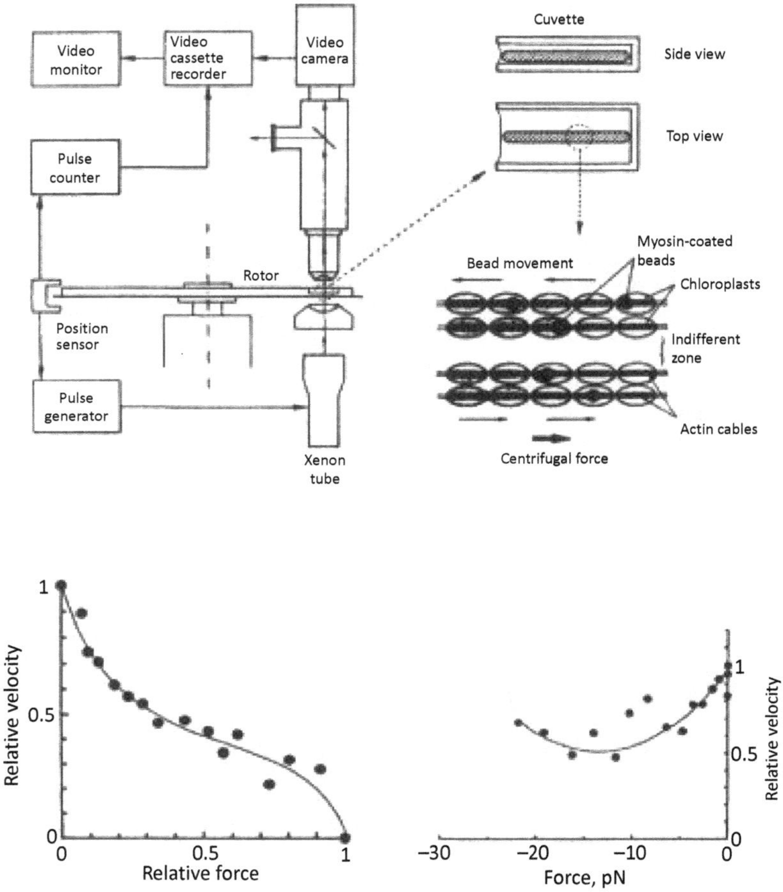

Figure 22 In vitro motility assay system combined with a centrifuge microscope. (A) Diagram of experimental arrangement. Myosin-coated beads are made to slide along actin cables in algal cell segment, placed in a cuvette on the rotor of centrifuge microscope. Constant centrifugal forces are applied to moving beads as constant external loads. (B) Steady-state force–velocity (*P*–*V*) relation for positive external loads, produced by centrifugal forces in the direction opposite to bead movement. (C) steady-state force–velocity relation for negative loads, produced by centrifugal forces in the same direction as that of bead movement. From Oiwa et al. (1990).

As illustrated in Fig. 22A, myosin-coated polystyrene beads were made to slide along actin cables in an algal cell segment, mounted in a cuvette on the rotor of a centrifuge microscope. By changing the velocity of rotor rotation, various constant centrifugal forces were applied to moving beads as constant external loads. The bead movement was observed under a microscope with a stroboscopic device and recorded on a video cassette for analysis. As shown in Fig. 18B, the relation between the amount of external load (=force, *P*) and the velocity (*V*) of ATP-dependent bead movement was found to be hyperbolic in shape in low load region, while it deviated from hyperbola to exhibit a distinct hump in the high-load region. Since it has been known in both single intact and skinned muscle fibers, distinct hump appears in the high-load region (Edman, 1988; Sugi et al., 1992), the shape of *P–V* relation of ATP-dependent in vitro actin–myosin sliding (Fig. 22B) indicates that a small number (probably <10) of myosin heads retain the basic characteristics of the contracting muscle. Meanwhile, the *P–V* relation for negative external loads (Fig. 22C) is currently a mystery for us.

Figure 23 Scanning electron micrographs of myosin-coated bead surface. Myosin filaments run in random directions. Arrows in the lower panel indicate myosin heads extending from myosin filaments. From Takahashi et al. (1993).

Using a high-power scanning microscope, we examined the manner of attachment of myosin on the bead surface (Takahashi et al., 1993). As shown in Fig. 23, myosin filaments were observed to run in random directions with myosin heads extending from them. In our impression, the ATP-dependent bead movement along actin cables is caused by myosin heads on myosin filaments, which are more or less parallel to actin filament arrays in actin cables. Strangely, none of the investigators using in vitro motility assay systems, including the optical trap device to be described later, performed electron microscopic observations on the manner of the attachment of myosin, HMM, or S-1 to the substratum, although the myosin head performance is greatly affected by the manner of their orientation and attachment on the substratum.

3.4 The Optical Trap System

Although it is possible to measure the force generated by the myosin heads fixed on a bead surface by recording the bending of a glass microneedle, connected to actin filament interacting with myosin (Kishino and Yanagida, 1988) (Fig. 24A), it is difficult to know the number of myosin heads involved in force generation to bend the needle. A much more sophisticated method to measure the force generated by a single myosin head (HMM or S-1) has been developed by Finer, Simmons, and Spudich (1994). This method utilizes an optical trap (or optical tweezers) to manipulate a single actin filament under microscopic observation. This technique is based on the power of a focused laser beam to draw and trap a small object, such as latex or polystyrene beads within the beam, and allows moving the trapped object by moving the position of the laser beam (Ashkin and Diziedzic, 1990) (Fig. 24B).

3.4.1 Advantages of the Optical Trap System

The optical trap method has the following advantages over other motility assay methods: (1) It does not require technical skill; (2) the trap stiffness can be easily changed by changing the laser beam intensity; and (3) the measurement of the force generated by a single myosin head under nearly isometric condition is possible by using a feedback circuit.

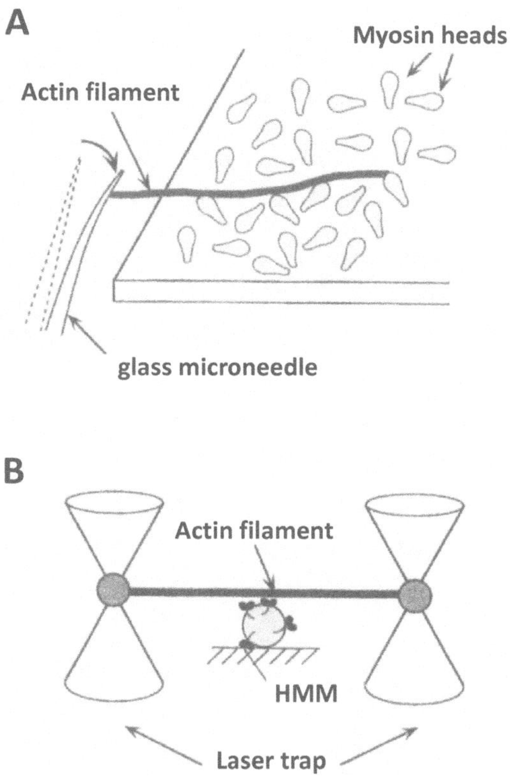

Figure 24 (A) Measurement of force generated by myosin heads by bending of a glass needle. (B) Optical trap device to measure force generated by a single myosin head (HMM). From Sugi (1998).

An example of records of force generation by a single myosin head in nearly isometric condition is shown in Fig. 25. It can be seen that force spikes of nearly uniform amplitude take place intermittently as a result of ATP-dependent interaction between a single myosin head and an actin filament. The duration of force spikes is variable, and sometimes very long (up to >1 s), while the interval between individual force spikes is also sometimes very long (> 2 s or more).

These features of force spike records result from low ATP concentration in the experimental solution (>1 μM), which is necessary to distinguish individual force spikes from each other. The intracellular ATP concentration in the muscle is ~2 mM, so that the attachment–detachment cycle between myosin heads and actin filaments goes on rapidly. If the ATP concentration is reduced to the micromolar range, individual myosin heads, which finished their power stroke to form rigor complex A-M, have to wait for the next ATP for a considerable time. In this situation, the time period in which individual myosin heads stay in the force-generating configuration may also be prolonged at the micromolar range of the ATP concentration. These aspects are important in interpreting the results of our electron microscopic studies to record the ATP-induced power stroke in individual myosin heads, which will be described later.

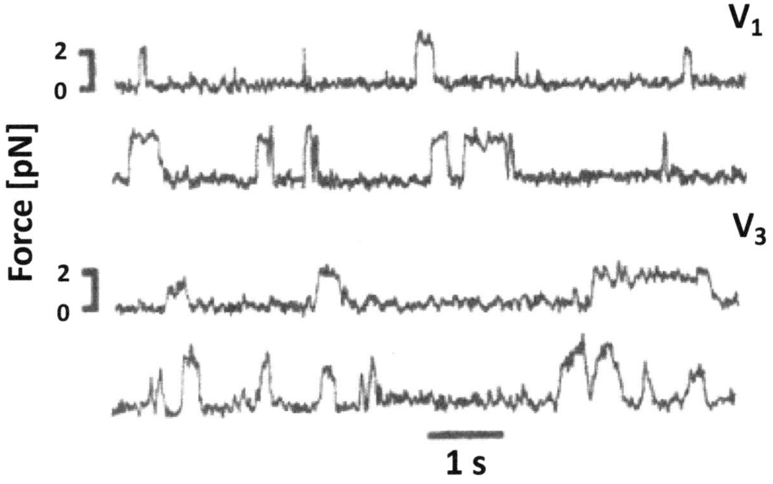

Figure 25 Records of force spikes obtained from two different rabbit cardiac muscle myosins V_1 and V_3 under a high optical trap stiffness. Note that long duration of force spikes as well as long intervals between force spikes. From Sugiura et al. (1998).

3.4.2 Shortcomings of the Optical Trap System

Using the optical trap method, many investigators measured the maximum force generated by a single full-length myosin, HMM, or S-1 with values ranging from 1.7 to 13 pN (Kaya and Higuchi, 2013). On the other hand, the unitary displacement (power stroke) caused by full-length myosin, HMM, or S-1 also shows a wide range of variation, ranging from 3 to 10 nm (Kaya and Higuchi, 2013). The marked diversity of the values seems to result mostly from uncertainties about the mode of fixation and orientation of myosin, HMM, or S-1 on the bead surface. In my opinion, therefore, the results obtained from the optical trap experiments bear no direct relation to what is actually taking place in muscle, though they are interesting in themselves.

Despite the above shortcomings of in vitro motility assay systems in providing correct information about the muscle contraction mechanism at the molecular level, many investigators still believe that the optical trap method is the most sophisticated, effective way to approach the contraction mechanism. As a result, sometimes papers on muscle fiber experiments are rejected by mainstream journals simply because the results are not consistent with the optical trap data. The excess belief in the in vitro motility assay systems thus tends to retard the progress of science and can be regarded as a kind of dogma to be broken for the progress of science.

Chapter 4

The Swinging Lever Arm Mechanism: A Most Prominent Current Dogma

Although the contraction model of Huxley–Simmons (1971) that force generation in the muscle results from the tilting of myosin heads (Fig. 8) was not supported experimentally, it is still believed that the unitary myofilament sliding is caused by some conformational changes of myosin heads attached to actin filaments, as predicted by H. E. Huxley (Fig. 4), a great pioneer who opened the current research field in muscle contraction. I will first explain the structure of the myosin head, which was determined crystallographically, since the current dogma of muscle contraction has been constructed on it.

4.1 Construction of the Swinging Lever Arm Hypothesis

Figure 26 shows the structure of the myosin head (myosin subfragment-1, S-1) of the vertebrate skeletal muscle, consisting of a catalytic domain (CAD) and a lever arm domain (LD), which are connected by a small converter domain (COD). Both the actin-binding and the ATP-binding sites are located in CAD,

Mysteries in Muscle Contraction: Evidence against Current Dogmas
Haruo Sugi

ISBN 978-981-4774-48-2 (Hardcover), 978-1-315-10191-0 (eBook)
www.panstanford.com

while LD is connected to the myosin filament backbone via myosin subfragment-2 (S-2, not shown).

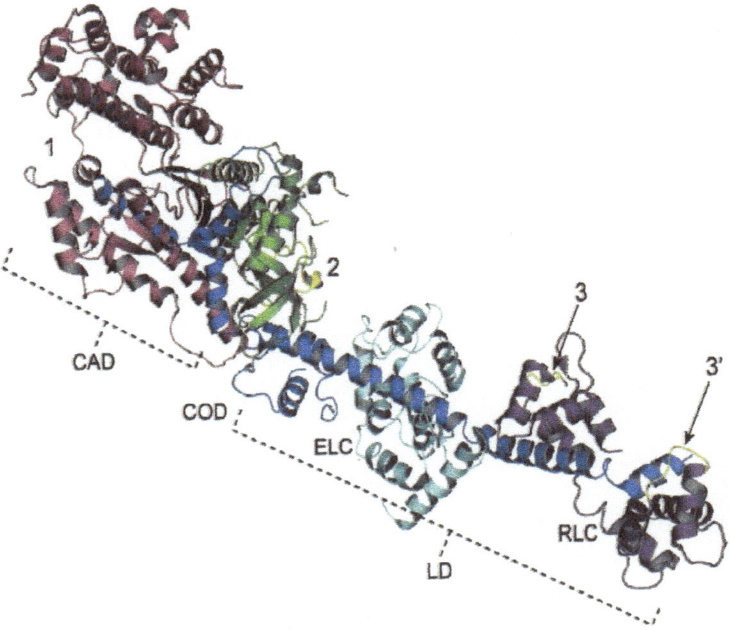

Figure 26 Myosin head structure. Catalytic domain (CAD) comprises 25 K (green), 50 K (red) and part of 20 K (dark blue) fragments of myosin heavy chain, while lever arm domain (LD) comprises the rest of 20 K fragment and essential (ELC, light blue) and regulatory (RLC, magenta) light chains. CAD and LD are connected with small COD. Location of peptides around Lys83 and that of two peptides (Met58~Ala70, and Leu106~Phe120) in LD are colored yellow. Approximate regions of attachment of antibodies 1, 2, and 3, used in our studies to be described later, are indicated by numbers 1, 2, and 3 and 3′, respectively. From Minoda et al. (2011).

To obtain information about the structural changes of myosin head, which is associated with myosin head power stroke, Rayment et al. (1993a,b) made extensive studies on nucleotide-dependent changes in myosin head crystal structure. In preparing the myosin head crystal, they used truncated myosin head obtained from slime mold (*Dictyostelium*) myosin, from which most of lever arm domain (LD)was detached except for its base, which remained attached to the converter domain. By removing most of the rod-shaped LD domain, the round-shaped truncated myosin head easily formed crystals to be used for experiments.

The reason why they used slime mold myosin instead of vertebrate skeletal muscle myosin was that myosin head crystals could be prepared more readily from the former than from the latter, though the truncated S-1 crystal structure was almost the same between them. At that time, I was rather surprised at the close similarity of S-1 crystals between the slime mold and the vertebrate animal despite an extreme difference in phylum between the two animals. It may be that due to close packing of S-1 molecules arising from the formation of crystals, differences in the S-1 structure between them would be forcibly eliminated.

Nevertheless, the nucleotide-dependent structural changes of *Dictyostelium* truncated S-1 have been studied intensively by many investigators, using a number of nucleotide analogs (Fisher et al., 1995; Geeves and Holmes, 2005; Sweeny and Houdusse, 2010). It was found that depending on the kind of nucleotide-analogs bound to S-1, the base of LD remaining in COD was observed to rotate around COD by 60°. This rotation of the remaining LD base was taken to represent pre- and post-power stroke states of the myosin head. Frankly speaking, this conclusion was too quick-minded and incomplete, since the observed nucleotide-dependent LD base rotation does not necessarily mean that non-truncated myosin heads connected to the myosin filament backbone via myosin head LD and S-2 actually rotate around CD; it seems rather unlikely that small structures within the small COD region can generate enough torque to rotate the massive COD around it, while the LD-S-2 junction moves only passively. It reminds me of an idiom, "the tail waggling the dog," a mechanism extremely unlikely in the myofilament lattice structure.

4.2 Shortcomings of the Swinging Lever Arm Hypothesis

Despite these uncertainties, however, the swinging lever arm hypothesis was constructed, in which the myosin head CAD, attached to the actin filament, moves without changing its attachment angle, as the result of the active rotation of LD around

COD. As illustrated in Fig. 23, the myosin head power stroke is produced by the active rotation of the CAD region around the COD region, which serves as a pivot. As the result, the myosin head CAD moves along the myosin filament long axis without changing its attachment angle with the actin filament. This manner of the myosin head CAD is consistent with the X-ray diffraction results that no appreciable tilting of CAD occurs during contraction and stretch. It should, however, be noted that the angle between myosin head LD and S-2 also changes during the myosin head power stroke. It seems strange, however, that in the swinging lever arm hypothesis, no attention is paid to the functional significance of angle changes at the LD-S-2 junction.

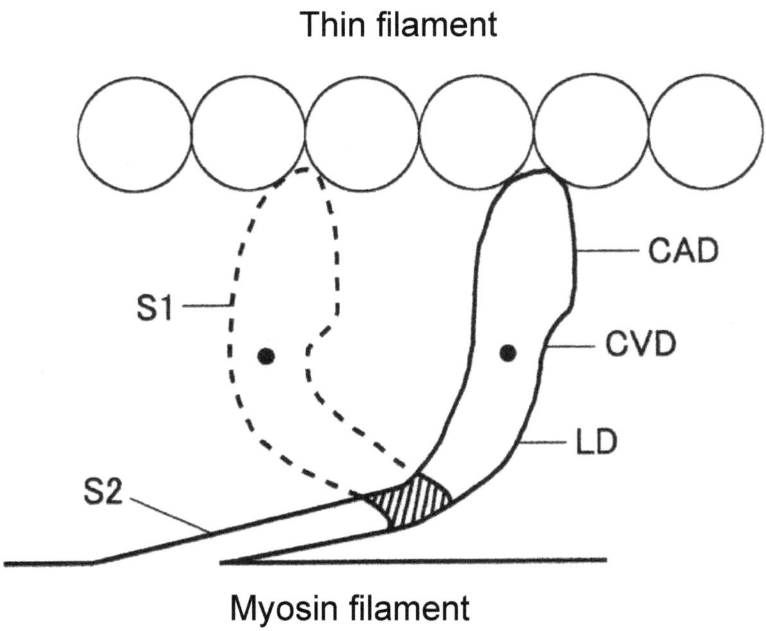

Figure 27 Myosin head configurations before (solid line) and after (broken line) power stroke, respectively. Myosin head LD rotates around COD (filled circles indicate points of pivot). Note that myosin head CAD attached to actin filament does not change its angle of attachment during the course of power stroke. Note also that the boundary (shaded area) between LD and S-2 serves also as a pivot for the myosin head motion.

As diagrammatically illustrated in Fig. 27, the myosin head CAD keeps its angle of attachment to the actin filament constant (without tilting) during the course of the power stroke, so that the CAD movement occurs parallel to the myosin filament long axis. The change in the myosin head configuration is assumed to be caused by the active rotation of myosin head LD around COD (filled circles indicate points of pivot), while the rotation of LD around the LD-S-2 junction (shaded area) is assumed to be passive. I find no reason to validate the ignorance of the possible functional role of the myosin head LD and myosin S-2 in producing the myosin head power stroke.

Despite its incompleteness (or biased character), the swinging lever arm hypothesis has been widely accepted by muscle investigators and now appears in many textbooks, probably because textbook authors like clear-cut stories to make their books attractive. The swinging lever arm hypothesis was constructed mainly by crystallographists without physiological considerations, so that the obviously important role of myosin LD and S-2, which support and transmit the myofilament sliding force to the myosin filament backbone, was ignored. This issue will also be discussed later. In fact, Andrew Huxley expressed his skepticism about this hypothesis (A. F. Huxley, 1998).

Chapter 5

Early Electron Microscopic Studies on Myosin Head Movement Using Quick-Freeze, Deep-Etch Replica Method

As described above, recent studies on the molecular mechanism of muscle contraction have been made mainly by in vitro motility assay experiments and crystallographic studies. Although the former method enables us to visualize actin–myosin sliding under the light microscope, it does not provide enough information to reach a full understanding of the contraction mechanism, because the experimental system differs too far from the 3D myofilament lattice structure in the muscle (Sugi et al., 2014). As a matter of fact, at a muscle symposium that I organized in 2004, Professor H. E. Huxley stated in his concluding remarks that the in vitro motility assay method did not make remarkable contribution to this research field, although it once excited people.

Crystallographic studies, on the other hand, are based on static crystal structures and cryo-electron micrographs obtained from extracted protein samples, ignoring possible important roles of myofilament structures, removed during the preparation of protein crystals and samples for cryo-electron microscopy. In order to study dynamic structural changes of myofilaments in the muscle during contraction, I used the quick-freeze, deep-etch

Mysteries in Muscle Contraction: Evidence against Current Dogmas
Haruo Sugi

ISBN 978-981-4774-48-2 (Hardcover), 978-1-315-10191-0 (eBook)
www.panstanford.com

replica method (Hauser and Cooke, 1983) combined with a digital image processor.

5.1 General Feature of the Freeze-Etch Replica Images

As shown in Fig. 28A, actin and myosin filaments as well as M-line and Z-band structures were clearly observed in freeze-etch replica images of quickly frozen mechanically skinned frog muscle fibers with the device illustrated in Fig. 11. The sarcomere length (~2.3 μM) was the same as that measured before freezing. High-magnification views of actin and myosin filaments in the A-band region, i.e., the region of the myofilament overlap, in relaxed, contracting, and rigor states, are shown in Fig. 28B–D, respectively. In all states, one actin filament was frequently seen between two myosin filaments, and they ran in parallel with each other; there was only a dark background in the gap between actin and myosin filaments. This indicates that in these replicas, only a single layer of myosin filaments and a single layer of adjacent actin filaments are exposed for the electron microscopic observation.

Stereo-view observations of replica images showed that actin filaments were located at the level slightly underneath myosin filament plane. Taking hexagonal myofilament lattice structure into consideration, we consider that in the present study, the myofilament lattice is fractured along the line shown in Fig. 29A. A myosin filament in vertebrate skeletal muscle consists of three parallel subfilaments (Mau and Rowe, 1980; Kensler and Stewart, 1983), and in each 14.3 nm-spaced axial level, three pairs of myosin heads are located on the myosin filament at angular intervals of 120° (Squire, 1972) (Fig. 29B). Myosin heads on the myosin filament therefore form three stranded helical arrays with a repeat of the structure at 42.9 nm intervals. Consequently, there should be ~12 pairs of myosin heads (or ~24 myosin heads) per 0.5 μM per one specific side of myosin filament (500/42.9 = 11.7). Assuming that each bridge-like structure between actin and myosin filaments corresponds to one myosin head, the proportion of myosin heads, observed as a bridge-like structure along one specific side of the myosin filament, can be estimated from the present results to be ~20%, 74%, and 70% of the total myosin heads in relaxed, contracting and rigor states, respectively.

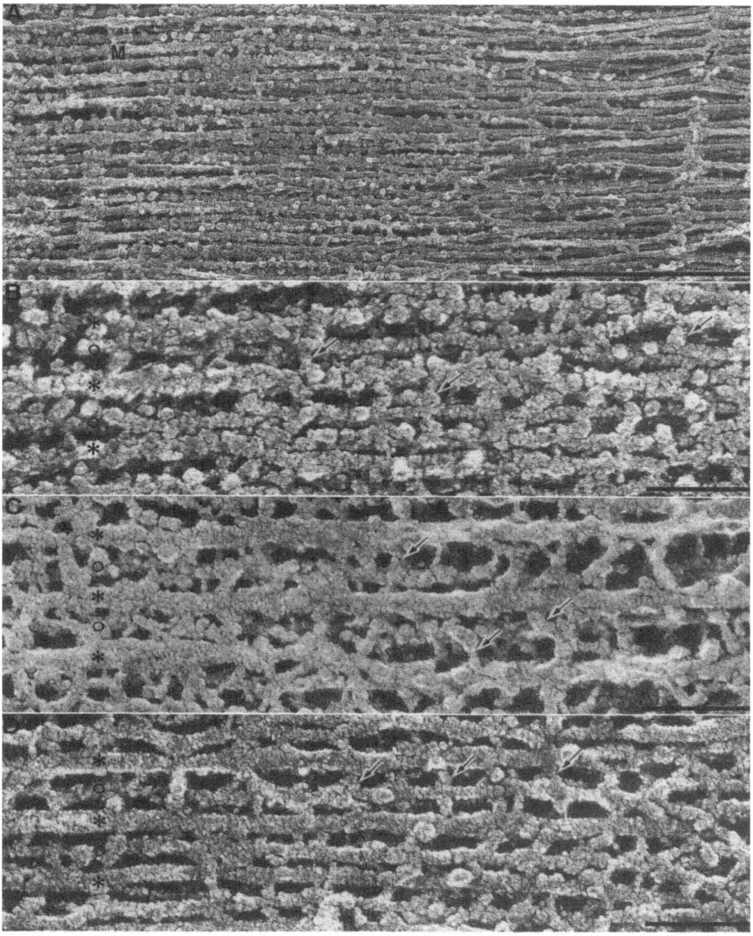

Figure 28 Freeze-etch replica images of skinned muscle fibers in various physiological states. (A) Low-magnification view of a half sarcomere in a relaxed fiber. M-band (M) and Z-band (Z) structures as well as actin and myosin filaments are clearly visible. Bar, 0.5 μM. (B–D) High-magnification view of actin and myosin filaments in relaxed (B), contracting (C) and rigor (D) states. Note that myosin (✳) and actin (O) filaments appear alternately and run in parallel with one another. Some of the rod-like myosin heads between adjacent myosin and actin filaments are indicated by arrows. Bar, 0.1 μM in B–D. From Suzuki et al. (1993).

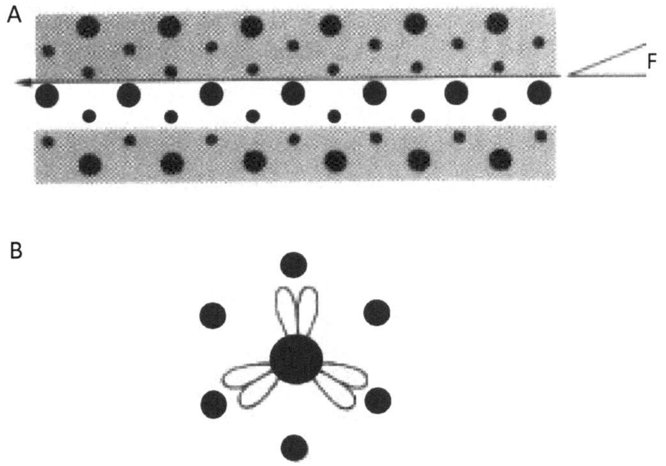

Figure 29 (A) Rationale of freeze-etch replica image in which actin and myosin filaments appear alternately. The freeze-fractures (F) in the present study travel through the fiber cross-section as indicated by arrow to remove myofilament lattice in the upper shaded area, so that subsequent deep-etch process exposes a layer of myosin filament (large filled circles) and a layer of actin filaments (small filled circles), while myofilament lattice in the lower shaded area remains embedded in ice. (B) Cross-sectional view of a myosin filament (large filled circle) at a 14.3 nm spaced axial level, where three pairs of myosin heads are located at angular intervals of 120°. Each myosin filament is surrounded by six neighboring actin filaments (small filled circles). From Suzuki et al. (1993).

In these studies, it is in principle not possible to distinguish whether myosin heads are actually attached to actin filaments or not, and therefore the results are not conclusive. Figure 30 shows histograms of the distribution of the axial interval between two adjacent myosin heads apparently attached to actin filaments. In the relaxed state, the interval of myosin heads, apparently attached to actin filaments, showed an extremely wide range of variation simply because a large proportion of myosin heads are located in the vicinity of myosin filaments and do not extend toward the actin filament (Fig. 30A). On the other hand, in both contracting and rigor states, the variation of the interval between the myosin heads attached to actin filaments was much smaller (Fig. 30B,C). Interestingly, in both states, the interval between two adjacent myosin heads, attached to the actin filament, showed a ~35 nm peak, since this value is similar to axial repeat of actin filaments as well as helical repeat of actin helix, determined by X-ray diffraction (Fig. 4C).

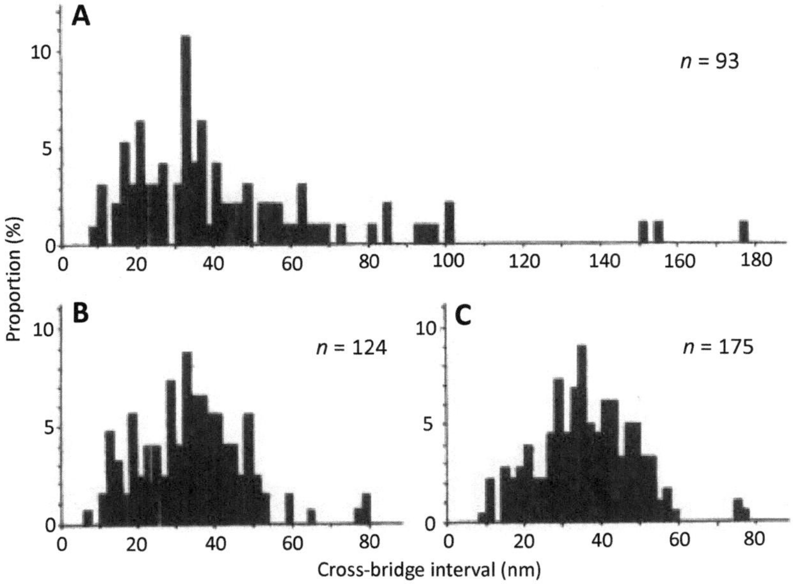

Figure 30 Histograms showing distribution of interval between two adjacent myosin heads attached to actin filaments, along each side of myosin filament in relaxed state (A), in contracting state (B) and in rigor state (C). Number of measurements is given alongside each histogram. From Suzuki et al. (1993).

5.2 Myosin Head Angle Distribution in Freeze-Etch Replica Images

The myosin head angles to myosin and actin filaments were measured on the electron micrographs of replica images by using a digital image processor (Suzuki et al., 1993), as illustrated in Fig. 31. As shown in Fig. 31E, the angle between a myosin head and myosin or actin filament was determined as the angle between diagonal lines of two rectangles, consisting of vertical and horizontal lines; one rectangle was drawn to just surround central axis of the whole myosin head, while the other rectangle was drawn to just surround a segment (length, 50~100 mm) of myosin or to both sides from the point of intersection of myosin head and actin or myosin filament axis. The myosin head angle was measured from the M-band side in each sarcomere; when a myosin head inclined toward the M-band, its angle to the myosin filament was acute, while its angle to the actin filament was obtuse.

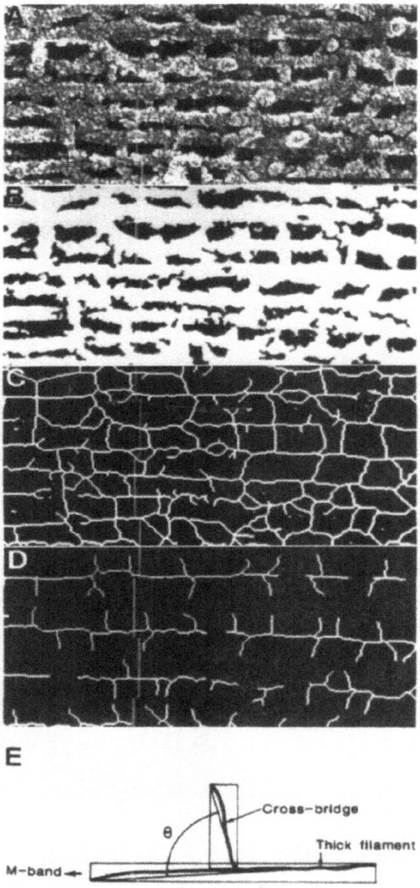

Figure 31 Method of measurement of myosin head angles to myosin filaments on freeze-replica images of quickly frozen muscle fibers. The original replica image of myofilaments and myosin heads (A) was first converted into binary image (B) and further conveyed into configurations (C) by thinning process. After erasing lines of actin filaments, pictures (D) only containing lines of myosin heads and actin and myosin filaments are obtained. Myosin head angles to actin or myosin filaments were measured as angle θ between two diagonal lines as shown in E. Bar, 0.1 μM (Suzuki et al., 1993).

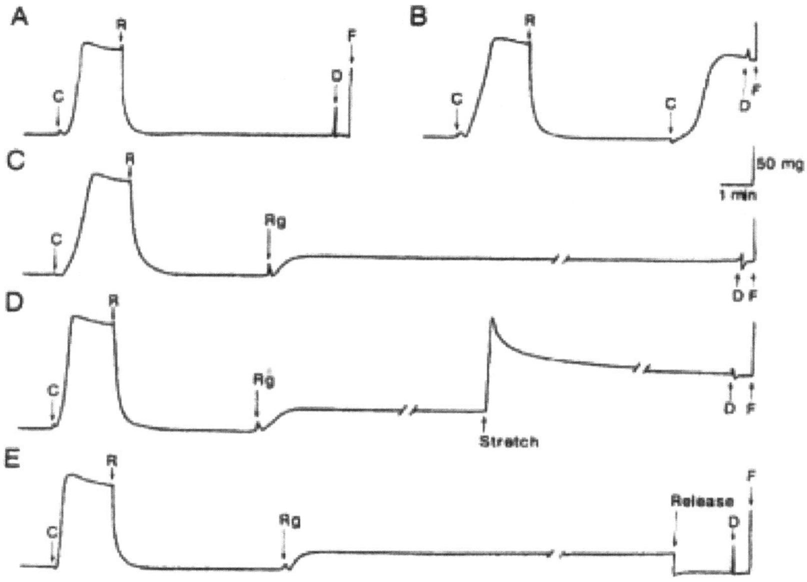

Figure 32 Force records when the fiber was quickly frozen in relaxed state (A), in Ca^{2+}-activated contracting state (B), in rigor state after removal of external ATP (C), in rigor state after stretch (D), and in rigor state after release to drop rigor force to zero (E). In each record, time of application of contracting, relaxing and rigor solution is indicated by C, R, and Rg, respectively, while time of draining experimental solution in the chamber and subsequent quick freezing is indicated by D and F, respectively. Suzuki et al. (1993).

The experimental procedures to quickly freeze single skinned fibers are shown in Fig. 32. In all cases, the fiber was first made to contract isometrically with contracting solution C (pCa = 4), and then made to relax with relaxing solution R (PCa > 9) to record the maximum isometric force P_0. The fibers in the relaxed state were frozen in the relaxing solution by draining the experimental solution in the chamber and then quickly frozen with a jet of liquid propane (A) (see Fig. 11). The fibers in the contracting state were frozen during isometric force generation in the contracting solution (B). The fibers in the rigor state were frozen during the generation of rigor force (amounting ~0.2 P_0) (C). Rigor fibers were also frozen during the generation of the increased rigor force (~0.6 P_0) by stretching them by ~5% (D), and during the generation of zero force by releasing them (E).

5.2.1 Myosin Head Angular Distribution in Relaxed and Contracting States

Histograms showing myosin head angular distribution to myosin and actin filaments in relaxed and contracting states are presented in Fig. 33. In the relaxed state, myosin head angles showed a wide range of variation from 30° to 150° with a peak at 90° to 95° for angles to the myosin filament, and peak at 85° to 90° for angles to the actin filament (Fig. 33A,B). The proportion of myosin heads taking angles of 90 ± 10° was 40.9% for angles to the myosin filament and 47.9% for angles to the actin filament (Fig. 33A,B).

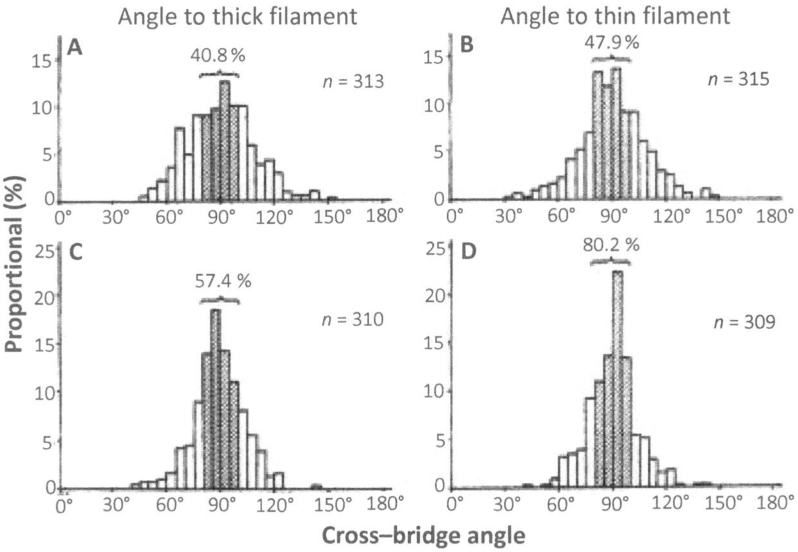

Figure 33 Histograms showing distribution of myosin head angle to myosin and actin filaments in relaxed state (A and B) and in contracting state (C and D). Total number of measurements is given alongside each histogram. In this and Fig. 30, myosin head angles to myosin and actin filament are shown on the left and the right side, respectively. Suzuki et al. (1993).

In contracting states, the proportion of myosin heads taking angles of 90±10° increased to 57.4% for angles to myosin filament and 60.2% for angles to actin filament, with corresponding decrease in the proportion of myosin heads taking angles other than 90±10° (Fig. 33C,D). As a result, the shape of the histogram in the contracted state became much sharper around the peak compared to that in the relaxed state.

In the relaxed state, most myosin heads are not actually attached to actin and myosin filaments. Consequently, the myosin head distribution in the relaxed state may largely reflect the angular distribution of freely moving myosin heads, being consistent with the concept of A. F. Huxley's contraction model (1957) that in the relaxed state, myosin heads fluctuate around the equilibrium position where the myosin head generates no force. This point will be discussed again in relation to our work with the gas environmental chamber (EC) to directly record the myosin head movement electron microscopically.

5.2.2 Myosin Head Angular Distribution in Rigor States

Myosin head angular distribution in the rigor state was nearly the same as that in the contracting state as shown in Fig. 34A,B. When rigor fibers are stretched before quick freezing, myosin head angular distribution to myosin filament showed no appreciable change, while its angular distribution to actin filament exhibited an appreciable decrease in the proportion of myosin heads taking angles 90±10° from 55.5% to 46.3% (Fig. 34C,D). This seems to suggest that if myosin heads attached to actin filaments are displaced in the direction opposite to fiber shortening, their base on myosin filaments is rigid, while their distal region attaching to the actin filament changes their angle of attachment in response to the stretch as illustrated in Fig. 35A,B. If myosin heads attached to actin filaments are released, on the other hand, the proportion of myosin heads with angles 90±10° decreased for both myosin filament (from 50.3% to 40.4%) and actin filament (from 55.5% to 44.2%) (Fig. 34B). This might result from the tilting of myosin head caused by elastic recoil of actin filament in myofilament nonoverlap region, i.e., in the I-band (Fig. 35A,C). Although the above explanation is at present putative, it seems consistent with the nonlinear stiffness of myosin heads attached to actin filaments, which has been reported by many authors (for a review, see Kaya and Higuchi, 2010).

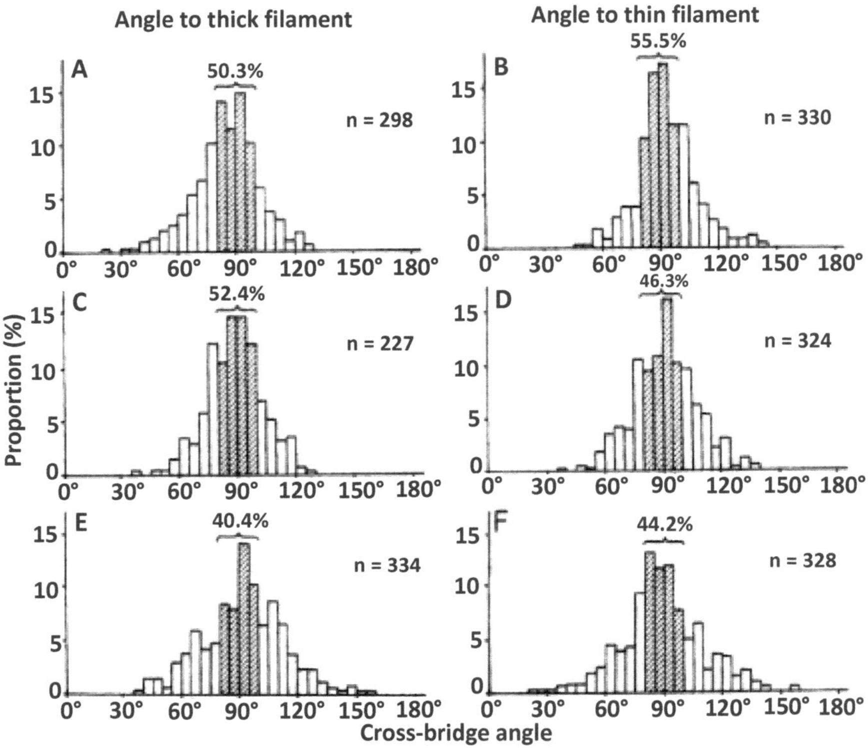

Figure 34 Histograms showing angular distribution of myosin heads to myosin and actin filaments in rigor state (A and B), in stretched rigor state (C and D), and in released rigor state (E and F). Suzuki et al. (1993).

To summarize, dynamic electron microscopy using the quick-freeze, deep-etch method gave us only indirect information about the behavior of myosin heads during muscle contraction. With the quick-freezing method, muscle fibers are frozen in ~1 ms. However, the myosin head configuration may not be kept the same as that at the moment of freezing; during the preparation of deep-etch replicas, myosin head structure might change to some extent. Such possible changes in the myosin head configuration cannot be excluded in principle, making the result ambiguous. This led me to perform electron microscopic studies to directly record myosin head movement associated with ATP hydrolysis, using the gas environmental chamber, which enables us to observe hydrated, living muscle myofilaments under high electron microscopic magnification. Fortunately, we could perform this type of experiment successfully with a number of novel results, as will be described in the next chapter.

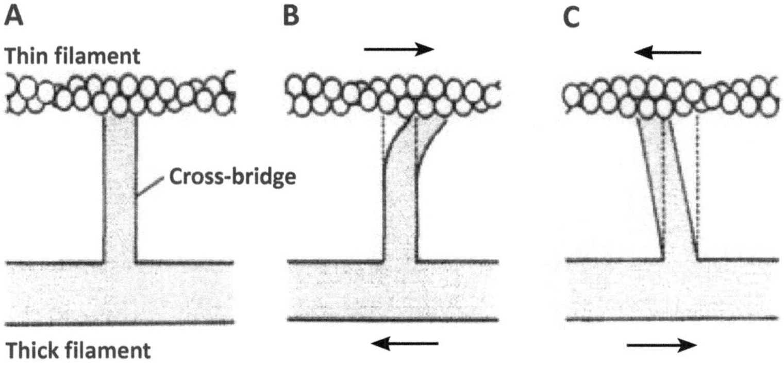

Figure 35 Possible changes in myosin head configuration attached to actin filament in response to stretch and release applied to rigor muscle fibers. (A) A myosin head with angle of 90° to both actin and myosin filament. (B) Stretch-induced bending of myosin head at its distal region. (C) Tilting of myosin head caused by releasing rigor muscle fibers. Suzuki et al. (1993).

Chapter 6

Electron Microscopic Recording of ATP-Induced Myosin Head Movement Using the Gas Environmental Chamber

The swinging lever arm hypothesis is a model based on the static structures of truncated myosin head crystals and acto-S1 complex, thus ignoring the possible role of myosin head LD and myosin-S-2, which were thrown away in preparing myosin head crystals and the acto-S1 complex. The validity of this model is still controversial even among muscle biochemists and biophysicists (e.g., Muretta et al., 2015). In the field of biochemistry, dynamic intramolecular structural changes can be studied by the method of fluorescence energy transfer (FRET) (e.g., Suzuki et al., 1998). Although this method provides information about the distance between two specific regions within a molecule, where fluorophores are attached, it can only be used for extracted protein samples in solution, a condition that differs too far from the 3D myofilament lattice structure. In addition, this method rests on the assumption that the attachment of fluorophores does not alter the protein structure and function at all. Our experiments to be described in this chapter were undertaken to make a breakthrough in this research field.

The most straightforward way to visualize and record the myosin head power stroke, which has been predicted by H. E. Huxley

Mysteries in Muscle Contraction: Evidence against Current Dogmas
Haruo Sugi

ISBN 978-981-4774-48-2 (Hardcover), 978-1-315-10191-0 (eBook)
www.panstanford.com

(Fig. 4) and still remains to be a matter for debate and speculation, is to perform experiments using the gas environmental chamber (or hydration chamber) attached to an electron microscope. The gas environmental chamber, which will hereafter be called "EC," enables us to study dynamic structural changes of living, hydrated biological macromolecules, like actin and myosin, in response to physiologically active substances such as ATP under an electron microscope.

In this chapter, I will briefly describe our efforts to establish the EC system suitable for the visualization and recording of ATP-induced myosin head movement, together with high evaluation of our EC experiments by two great pioneers, A. F. Huxley and H. E. Huxley, in contrast with the complete ignorance of our EC work by crystallographists and in vitro motility assay researchers.

6.1 The Gas Environmental Chamber

The gas environmental chamber (EC) was developed to observe physical and chemical reactions in solution under an electron microscope (Butler and Hale, 1981). Although the EC system is widely used by materials scientists for in situ observation and recording of chemical reactions, its application to life science was unsuccessful for many years.

In the early 1980s, great luckiness struck me to meet Professor Akira Fukami (Nihon University), a pioneer who developed an EC system suitable for biological research and was looking for a coworker. We started working together, using the EC system to visualize and record the myosin head movement. As shown in Fig. 36, the EC is a circular-shaped metal compartment (diameter, 35 mm; depth, 0.8 mm) with upper and lower window frames (copper grids) to pass electron beam. Each window frame has nine apertures (diameter, 0.9 mm), and covered with carbon sealing film, insulating the interior of the EC from high vacuum of electron microscope. The specimen is placed onto the surface of lower sealing film, and covered with thin layer of experimental solution (thickness ~200 nm), which is in equilibrium with circulating water vapor at a rate of 0.1~0.2 ml/min through the EC. To obtain clear specimen image, pressure in the EC is kept at 60~80 Torr. The EC is attached to a 200 kV transmission electron microscope (JEOL, JEM 2000EX). The EC contains a glass capillary microelectrode to apply ATP iontophoretically to the specimen.

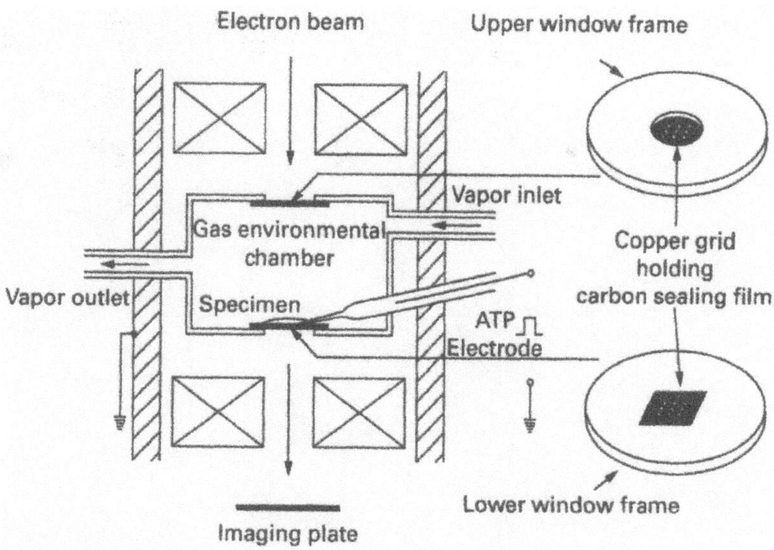

Figure 36 A gas environmental chamber (EC). Upper and lower windows of the EC are covered with carbon sealing film held on copper grids with nine apertures. Interior of the EC is constantly circulated with water vapor to keep biological specimen in wet, living state. For further explanation, see text. From Sugi et al. (1997).

6.1.1 Carbon Sealing Film

Both spatial resolution and contrast of the specimen image obtained using the EC increase with decreasing thickness of sealing film. Preliminary experiments made in Professor Fukami's laboratory indicated that the thickness of the sealing film should be 15~20 nm to obtain spatial resolution <1 nm. Meanwhile, the resistivity of the sealing film against the pressure difference decreases sharply with increasing area to be covered by the sealing film. After a number of trials and errors, Professor Fukami and Adachi (1965) solved this problem by using plastic microgrids made from cellurose acetobutylate, supporting the carbon sealing film (Fig. 37). Microgrids with fairly uniform holes of 5~8 μm diameter (Fig. 37, right) were suitable for the electron microscopic observation and recording of the specimen, i.e., synthetic myosin filaments, while those with too small or irregular sizes (Fig. 37, left and middle) were unsuitable. Sealing films made from silicon-nitrogen compound are now commercially available, though their spatial resolution seems not as good as that of carbon sealing film.

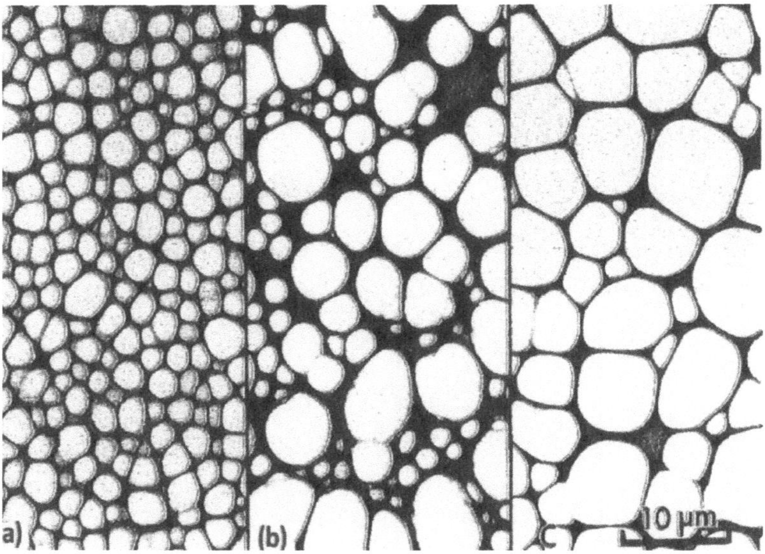

Figure 37 Light micrographs of carbon sealing film with too small holes (a), with holes of irregular size (b), and with holes of appropriate size for EC recording of specimen image (c). From Fukushima (1987).

6.1.2 Iontophoretic Application of ATP

A most characteristic feature of our EC is that it contains a glass capillary microelectrode, filled with 100 mM ATP (Fig. 36). Positive DC voltage is constantly applied to the electrode to prevent diffusion of ATP from the electrode. At any desired time, negative-going current pulses (intensity, 10 nA; duration up to 1 s) are applied to the electrode through a current clamp circuit (Oiwa et al., 1991, 1993). In each iontophoretic ATP application, about 10^{-14} mol ATP is released from the electrode and reaches the specimen by diffusion. The peak ATP concentration around the specimen is calculated to be 5~10 μM (Sugi et al., 1997). As explained later, the small concentration of ATP around the myosin head enabled us to record distinct changes in the position of individual myosin heads before and after completion of ATP-induced movement.

6.1.3 Critical Electron Dose Not to Impair Contractile Function of Myofilaments

To investigate physiological function of actin and myosin filaments mounted in the EC, it is essential to determine the critical incident electron dose not to impair contraction of myofibrils, i.e., small bundle of actin and myosin filaments isolated from the skeletal muscle, in which the regular arrangement of filaments (Fig. 2) is preserved.

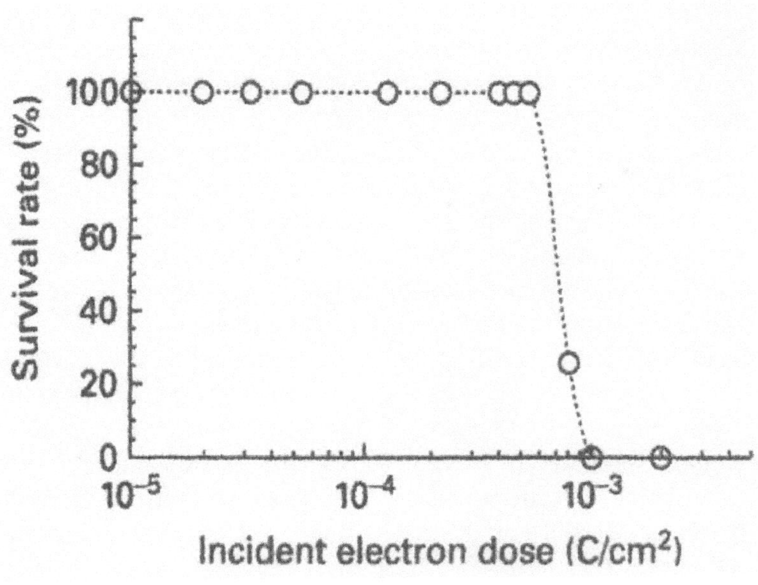

Figure 38 Relation between the total incident electron dose and the survival rate of myofibrils in the EC, representing percentage of myofibrils within an electron microscopic field to shorten in response to ATP. From Suda et al. (1992).

For this purpose, Suda, Ishikawa, and Fukami (1992) mounted isolated myofibrils, i.e., bundles of myofilaments, in the EC, and observed their contraction in response to ATP under low electron microscopic magnification (2,500×). As shown in Fig. 38, all myofibrils within a microscopic field shortened in response to ATP when the total incident electron dose did not exceed 5×10^{-4} C/cm^2. If the total incident electron dose was further increased above this critical value, ATP-induced shortening of myofibrils disappeared in a nearly all-or-none manner. Based on this result, electron microscopic observation and recording of the specimen in the EC was made within a limited time period, in which total incident electron dose did not exceed the critical value to impair physiological function of actin and myosin filaments. Up to the present time, most electron dose should be used to find samples in an electron microscopic field under low magnifications.

6.1.4 Position Marking of Individual Myosin Heads

Since myosin filament samples are mounted in the EC in unstained conditions, it is necessary to position-mark individual myosin heads to observe and record their ATP-induced movement. This was achieved by the use of site-directed monoclonal antibodies to various regions of myosin heads. We used three different antibodies. Two antibodies are prepared by Sutoh, Tokunaga, and Wakabayashi (1989) for functional mapping of the myosin head; one antibody attaches to the junctional peptide between 50 and 20 K fragments of the myosin heavy chain, located in myosin head CAD (antibody 1 or anti-CAD antibody), while the other antibody attaches to reactive lysine residue (Lys83) located in myosin head COD (antibody 2 or anti-COD antibody). Another antibody, prepared by us, attaches to both essential and regulatory light chains in myosin head LD (antibody 3 or anti-LD antibody). It can be seen in Fig. 39 that antibodies 1 and 2 attach around the distal region of myosin head (A and B), while antibody 3 attaches to the neck region of myosin head, where myosin head LD is located.

Using these antibodies, we could successfully position-mark individual myosin heads by (1) mixing a solution containing myosin filaments with a solution containing antibody to attach antibody molecules to their epitopes on myosin filaments, and then further mixing a solution containing myosin–antibody (IgG) complex with a solution containing gold particles (diameter, 15 or 20 nm, coated with protein A, serving as a paste, EY laboratories).

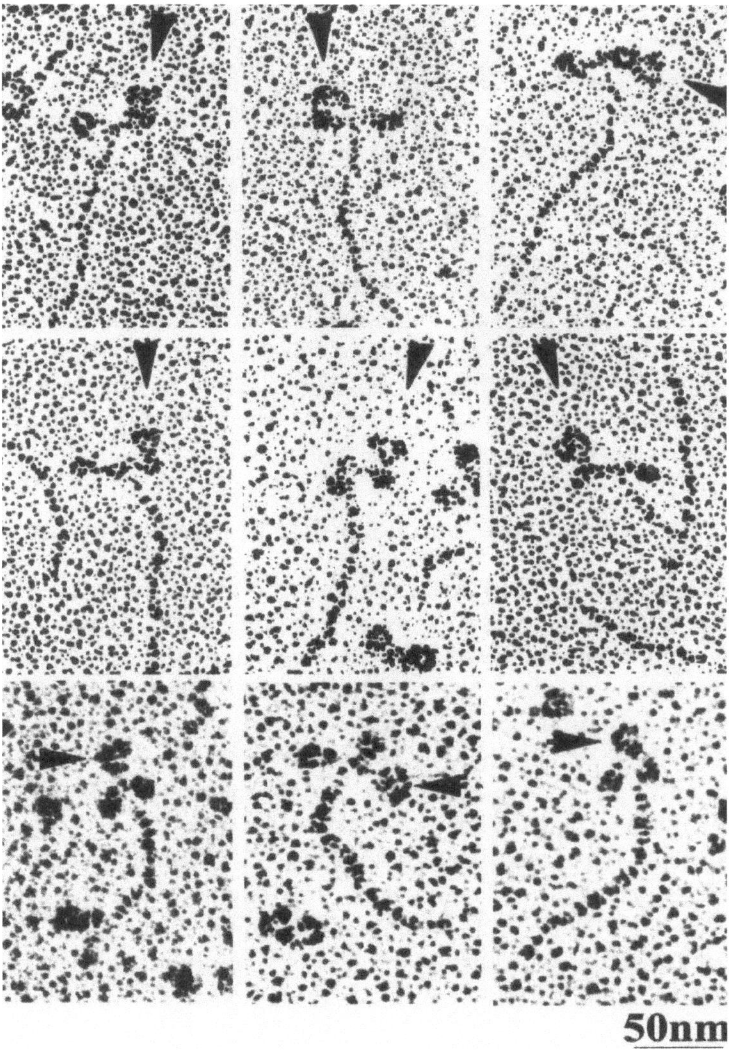

Figure 39 Galleries of electron microscopic images of rotary shadowed antibody 1- (upper panels), antibody 2- (middle panels), and antibody 3- (lower panels) myosin complex. Antibody (IgG) molecules attached to myosin are indicated by arrowheads. Upper and middle panels A are provided from the published paper (Sutoh, Tokunaga, and Wakabayashi, 1989).

6.2 Preliminary Experiments on Myosin–Paramyosin Core Complex in the Absence of Actin Filaments

As the first step of our EC experiments to visualize and record myosin head movement coupled with ATP hydrolysis, we used synthetic filaments, consisting of rabbit skeletal muscle myosin and thick molluscan paramyosin filaments, i.e., myosin–paramyosin core complex, because this material was much easier to handle compared to vertebrate myosin filaments which tended to curl and aggregate. The synthetic filaments, i.e., myosin–paramyosin core complex consists of long and thick paramyosin filaments (diameter, 50–200 nm; length, 10–30 μM), prepared from molluscan smooth muscle, and rabbit skeletal muscle myosin bound around them. We performed the following preliminary experiments in the absence of actin filaments.

6.2.1 Observation and Recording of Filament Image

Because of the absolute limitations to finish EC experiments within the total incident electron dose of 5×10^{-4} C/cm^2, we had to find, focus, and record the specimen with extremely weak electron beam intensities $< 5 \times 10^{-13}$ A/cm^2. Therefore, observation and focusing of synthetic filaments in the dark microscopic field required enormous technical skill and patience. The actual beam intensity through the specimen under an electron microscopic magnification of 10,000× was $5 \times 10^{-13} \times (10{,}000)^2 = 5 \times 10^{-5}$ A/cm^2. As soon as gold particles located on the specimen were focused, the electron beam was stopped until the time of recording. The exposure time of the specimen image to the IP was 0.1 s with a beam intensity of $1{\sim}2 \times 10^{-12}$ A/cm^2. In this condition, recording of the same specimen image could be repeated up to four times.

Figure 40 shows typical electron micrographs of the myosin–paramyosin core complex filaments, with a number of myosin heads, effectively position-marked with gold particles via antibody 1. Gold particles (diameter, 15 nm) attached to myosin heads were clearly observed as discrete dark spots, which were clearly distinguished from the background (Fig. 40A,C). As shown in Fig. 40B, the filaments were actually covered by myosin heads with gold particles attached.

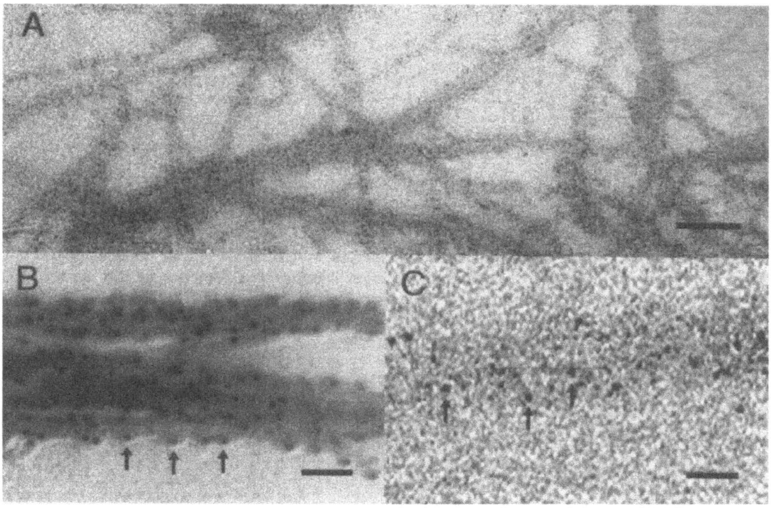

Figure 40 (A) Imaging plate IP) record of synthetic filaments with numerous gold particles attached to myosin heads. (B) Conventional rotary-shadowed electron micrograph of the filament. (C) Enlarged IP record of the filament. Bar, 500 nm in A, and 100 nm in B and C). Arrows indicate gold particles From Sugi et al. (1997).

6.2.2 Data Analysis

Filament images (magnification, 10,000×) were analyzed with an image processor. The pixel size was 5 × 5 nm in the preliminary experiments described in this section, and 2.5 × 2.5 nm in the later experiments on vertebrate muscle myosin filaments to be described later. Reflecting a limited number of electrons reaching a single pixel, the image of a gold particle consisted of 10 to 50 pixels. The center of mass position for each particle was determined as the coordinates (two significant figures) within a single pixel, where the center of mass position was located. These coordinates representing the position of the gold particles (and therefore the position of individual myosin heads) were compared between the two imaging plate records. The absolute coordinates common to the two records were obtained from the position of natural markers (bright spots on the carbon sealing film, which can be seen at the left side bottom of Fig. 40A). The distance (D) between the two center of mass positions, with the coordinates x_1, y_1, and x_2, y_2, respectively, was calculated as

$$D = \sqrt{(x_1 - x_2)^2 + (y_1 - y_2)^2}.$$

6.2.3 Stability in Time-Averaged Position of Myosin Heads

To examine whether the position of each gold particle remained unchanged with time in the absence of ATP, we took a pair of IP records of the same filament (magnification, 10,000×) at an interval of several min. Then we determined the center of mass position of each gold particle as the coordinates (two significant figures) within a single pixel (5 × 5 nm), and these coordinates representing the particle (and myosin head) position were compared between the two records. A typical result is shown in Fig. 41A, in which a circle of 15 nm diameter was drawn around each center of mass position. It was found that the position of each gold particle (and therefore the position of each myosin head) remained almost unchanged with time, if the limit of spatial resolution determined by the pixel size (5 × 5 nm) was taken into consideration. Figure 41B is a histogram showing the distribution of distance (D) between the two recorded positions of the same particle. Among 141 particles examined, 87 particles showed no significant position changes ($D < 5$ nm), and 50 particles showed only small position changes (5 nm $< D >$ 10 nm). These results indicate that though the position of each myosin head is expected to fluctuate due to thermal motion, its mean position time-averaged over 0.1 s (exposure time of imaging plate system) does not change appreciably with time in the absence of ATP (Sugi et al., 1997).

The stability in the position of gold particles (and therefore myosin heads) indicates that (1) the filaments are firmly fixed in position to the carbon sealing film, (2) the connections between myosin heads and antibody molecules (IgG) and the connections between antibody molecules and gold particles are rigid enough, so that no appreciable time-averaged fluctuation in the gold particle position can be recorded in the absence of ATP, and (3) as the result, the position of gold particles, time-averaged over 0.1 s is kept almost constant with time to serve as a good position-marker of myosin heads. As already mentioned, the stability of myosin head position found in our EC experiments is completely in accord with the contraction model of A. F. Huxley (1957), in which each myosin head undergoes thermal motion around a definite equilibrium position where the myosin head exerts no force.

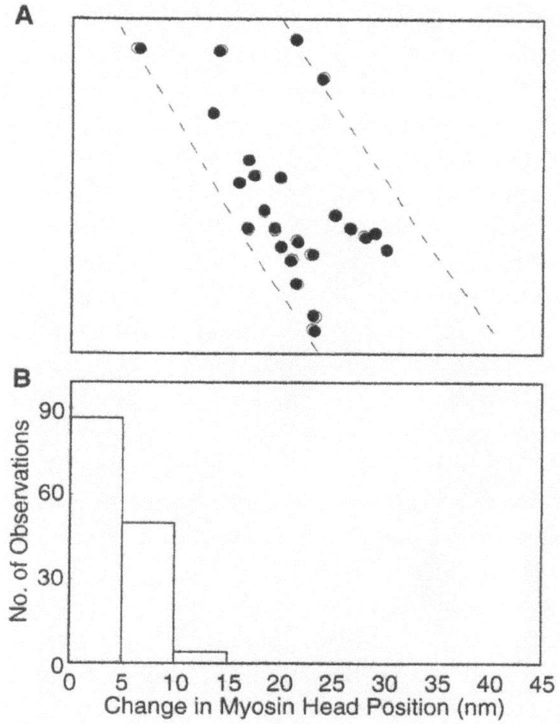

Figure 41 Stability in the position of myosin heads in the absence of ATP. (A) Comparison of the myosin head positions between the two records of the same filament on the common coordinates. Filled and open circles (diameter, 15 nm) are drawn around the center of mass position of each gold particle in the first and the second records, respectively. (B) Histogram showing the distribution of distance between the two center of mass positions of the same gold particle in the first and the second records. From Sugi et al. (1997).

We were very much encouraged with the constant time-averaged myosin head position, because it indicated that our attempt to record ATP-induced myosin head movement was promising.

6.2.4 ATP-Induced Myosin Head Movement

We found that in response to iontophoretically applied ATP, myosin heads moved along the filament long axis by ~20 nm (Fig. 42) (Sugi et al., 1997). This value was of course much larger than the expected amplitude of myosin head movement in vertebrate muscle myosin filaments, indicating that, in this particular synthetic filaments, myosin S-2 region shortens in response to ATP because of relatively loose attachment of myosin molecule to paramyosin core. The possibility of S-2 shortening in the contracting vertebrate skeletal muscle has been suggested by Harrington and his coworkers, which will also be mentioned later in this book. Since the experiments were performed in the absence of actin filaments, the recorded myosin head movement was regarded as the recovery stroke. Unfortunately, we failed to observe the myosin head movement across the filament bare region to make this point clear, because the filaments were too long and aggregated each other. On binding with ATP, myosin heads (M) form M-ADP-Pi, having average long life time >10 s, and this is the reason why we could record ATP-induced myosin head movement despite the limited time resolution of 0.1 s (exposure time for IP recording).

When we just succeeded in recording the ATP-induced myosin head movement in synthetic filaments, Professor A. F. Huxley stayed in Japan to participate in an international symposium concerned with the in vitro motility assay. During this meeting, he telephoned me, "*I very much miss you. Why you do not participate in this meeting?*" I answered, "*I am now disappointed with in vitro motility assay people in Japan, and am busy with performing the EC experiments with exciting results.*" After the symposium, he stayed at my laboratory for a few days and gave many instructive comments on our EC experiments. On leaving my laboratory, he said, "*Your experimental system using the EC is much closer to living muscle than in vitro motility assay system, I very much look forward to having your results on vertebrate myosin filaments.*"

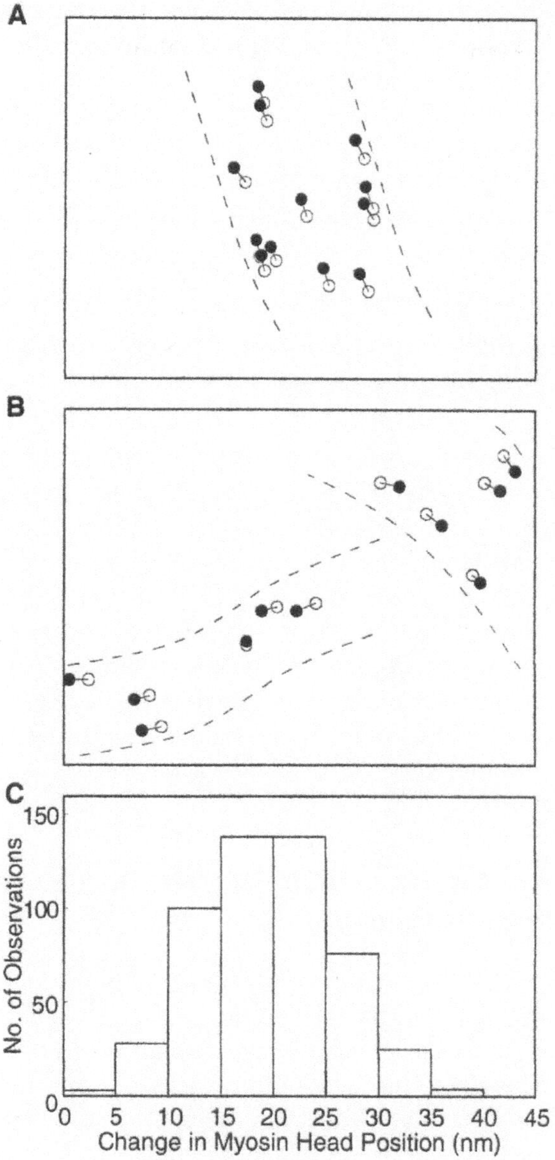

Figure 42 (A and B) Examples of myosin head position changes. Filled and open circles (diameter, 15 nm) indicated myosin head positions before and after ATP application, respectively. (C) Histogram of amplitude distribution of ATP-induced myosin head movement. From Sugi et al. (1997).

By the way, I also had an opportunity of talking about our preliminary success in recording myosin head movement in the myosin–paramyosin core complex at a symposium organized by Professor H. E. Huxley in Glasgow. After my talk, he said, "*Congratulations for your success! Please do experiments on vertebrate myosin filaments. Whatever the result may be, I will believe it as the amplitude of myosin head power stroke.*"

6.3 Experiments on Synthetic Vertebrate Muscle Myosin Filaments in the Absence of Actin Filaments

Here, I will describe our experimental results obtained from synthetic vertebrate muscle myosin filaments, which are long and thick and suitable for our EC experiments. I will also mention negative response of muscle crystallographists and in vitro motility assay researchers against our EC work. The recent appearance of many access-free journals obviously reflects the unfair behavior of people in every field of science to destroy publication of original work by those, whom they regard to be their competitors. In this connection, access-free journals protect the sound progress of science against unfair investigators.

6.3.1 Preparation of Vertebrate Muscle Bipolar Myosin Filaments

To study the ATP-induced myosin head movement in vertebrate muscle myosin filaments, the specimen used was synthetic myosin filaments, prepared by mixing myosin and myosin rod at a ratio of 1:1. Therefore, the synthetic myosin filament was myosin-myosin rod cofilament. Although the synthetic myosin filaments tended to aggregate, we could occasionally find spindle-shaped synthetic filaments with a number of gold particles on them. As shown in Fig. 43, each gold particle image was composed of 20 to 50 dark pixels with a wide range of gradation. The spindle-shaped synthetic filaments with two tapered ends indicated that the polarity of myosin heads extending from myosin filaments may be reversed across the filament center. This was actually confirmed by our success in recording the ATP-induced myosin head recovery stroke.

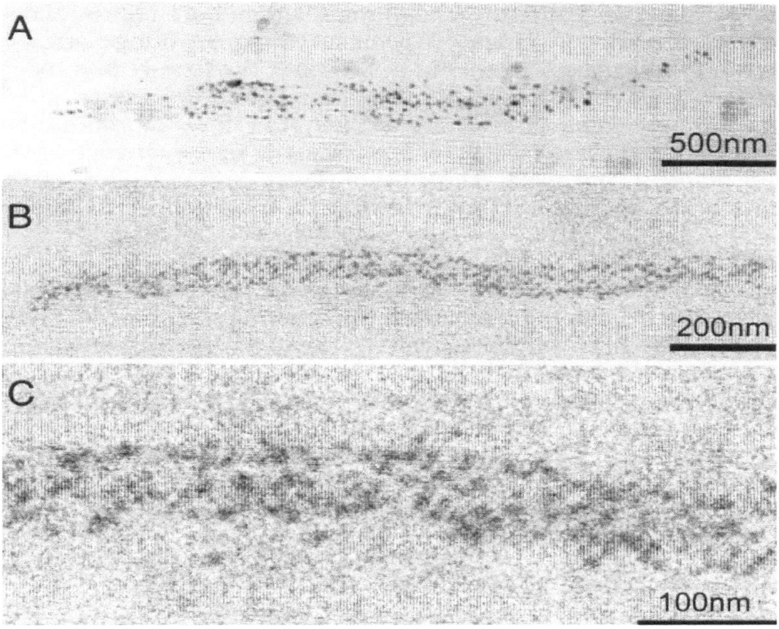

Figure 43 (A and B) IP records of spindle-shaped bipolar myosin filaments with gold particles attached to myosin heads. (C) Enlarged view showing part of the myosin filament shown in (B). From Sugi et al. (2008).

6.3.2 Stability in the Time-Averaged Position of Myosin Heads in the Absence of ATP

At the start of the experiments, we also examined the stability of the myosin head position, by taking two IP records of the same filament at intervals of several minutes. The results are shown in Fig. 44. As has been the case in the preliminary experiments on myosin–paramyosin rod complex filaments (Sugi et al., 1997), the time-averaged position of each gold particle, i.e., the position of each myosin head on myosin filaments, was found not to change appreciably with time, if the limit of accuracy of spatial resolution determined by the pixel size of the IP (2.5 × 2.5 nm; magnification,

10000×) was taken into consideration. As already mentioned, the stability of time-averaged position of individual myosin heads in the absence of actin filaments provides a favorable condition to record the ATP-induced myosin head movement.

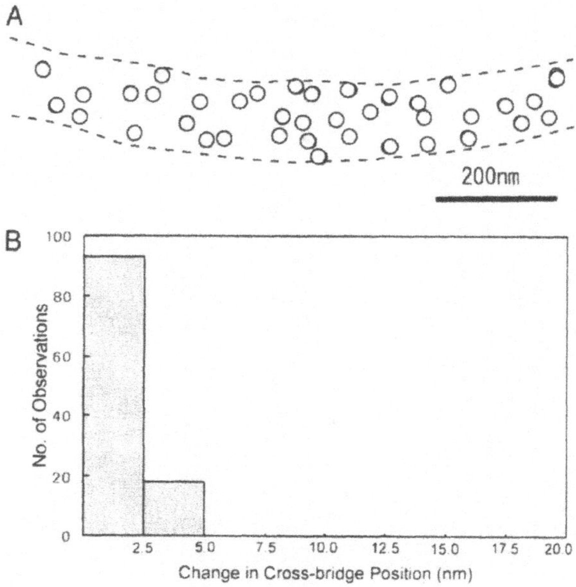

Figure 44 (A) Comparison of myosin head position between two IP records of the same filament. Open and filled circles of 20 nm diameter are drawn around center of mass positions of the first and the second IP records, respectively. Note that filled circles are almost invisible because of almost complete overlap between open and filled circles. (B) Histogram of distribution of distance between the center of mass position of gold particles between the two IP records. From Sugi et al. (2008).

The stability in the time-averaged myosin head position implies that both the myosin head–antibody and the antibody–gold particle linkages are rigid and stable, so that no appreciable "amplification" of myosin head movement occurs during EC experiments. As will be described later, the ATP-induced myosin head movement takes place in such a way that myosin head CAD moves parallel to myosin filament axis (Fig. 26, Minoda et al., 2011), precluding the possibility of "marked amplification" of myosin head movement, which may arise if the myosin head movement is rotational around the CAD-LD or around the LD-S-2 junction.

6.3.3 ATP-Induced Unidirectional Myosin Head Movement in the Absence of Actin Filaments

In response to ATP, individual myosin heads on the filaments mostly exhibited distinct unidirectional movement parallel to the filament long axis. The results obtained are summarized in Fig. 45. As can be seen in Fig. 45A, all myosin heads on the filament were found to move in one direction (from left to right) in response to ATP. The direction of the myosin head movement was parallel to the filament long axis. The histogram of amplitude distribution of the myosin head movement (Fig. 45B) exhibited a peak at 5~7.5 nm. The average amplitude of myosin head movement (excluding values < 2.5 nm) was 6.5 ±3.7 nm (mean ± SD, n = 1210).

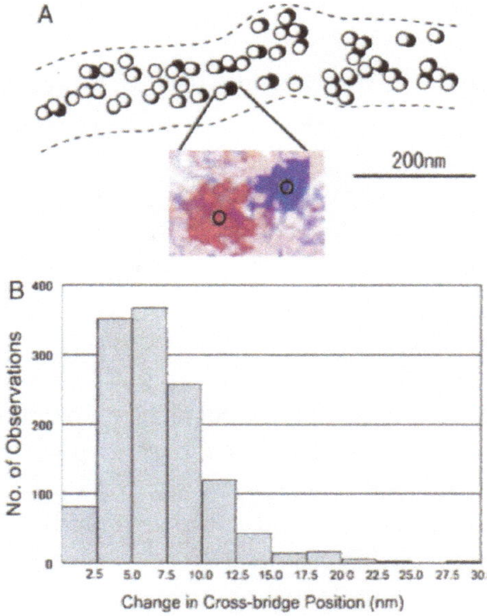

Figure 45 (A) Changes of the myosin head position between the two IP records, one taken before and the other taken after, application of ATP. Open and filled circles (diameter, 20 nm) are drawn around the center of mass positions of the same gold particle before and after ATP application, respectively. (Inset) Superimposed IP records showing the change in position of the same gold particle, colored red (before ATP application) and blue (after ATP application). The center of mass position is shown by small circle. (B) Histogram of amplitude distribution of ATP-induced myosin head movement. From Sugi et al. (2008).

In our experimental condition, gold particles were located on both upper and lower surfaces of the myosin filament, and they were equally in focus in the electron microscopic field. While the myosin heads located on the upper filament surface may move freely in response to ATP, the myosin heads located on the lower side of the filament would attach to carbon sealing film, and their movement might be more or less inhibited or reduced in amplitude. This possibility seems to be consistent with the frequent presence of gold particles, that did not move appreciably or moved with small amplitude < 5 nm. If this explanation is correct, the actual amplitude of freely moving myosin heads would be > 7.5 nm. Meanwhile, the ATP-induced myosin head movement disappeared if the myosin head ATPase activity was eliminated by N-ethylmaleimide. Myosin heads did not move in response to ADP.

6.3.4 Reversal in the Direction of ATP-Induced Myosin Head Movement Across the Bare Region

Despite great difficulties in finding gold particles located at both sides of the myosin filament bare region, across which the polarities of myosin heads are reversed, we fortunately succeeded in recording the ATP-induced myosin head movement, which were actually reversed in direction across the bare region. The results obtained are presented in Fig. 46. It can be seen that on application of ATP, individual myosin heads at both sides of the bare region move away from the bare region, but not toward it. Since actin filaments are absent in this experimental system, the ATP-induced myosin head movement at both sides of bare region is regarded as the recovery stroke, associated with reaction in the Lymn–Taylor scheme, M + ATP $\rightarrow$ M-ATP $\rightarrow$ M-ADP-Pi. As already mentioned, the complex M-ADP-Pi has a long lifetime > 10 s in the absence of actin, and this is the reason why the change in myosin head position after ATP application can be recorded with the limited time resolution of our experiments (exposure time of filament image to the IP, 0.1 s).

In principle, myosin head power and recovery strokes should be the same in amplitude and opposite in direction. Consequently, our results imply that the power stroke amplitude of myosin head is ~6.5 nm in this experimental condition, in which a large proportion of myosin heads move more or less freely.

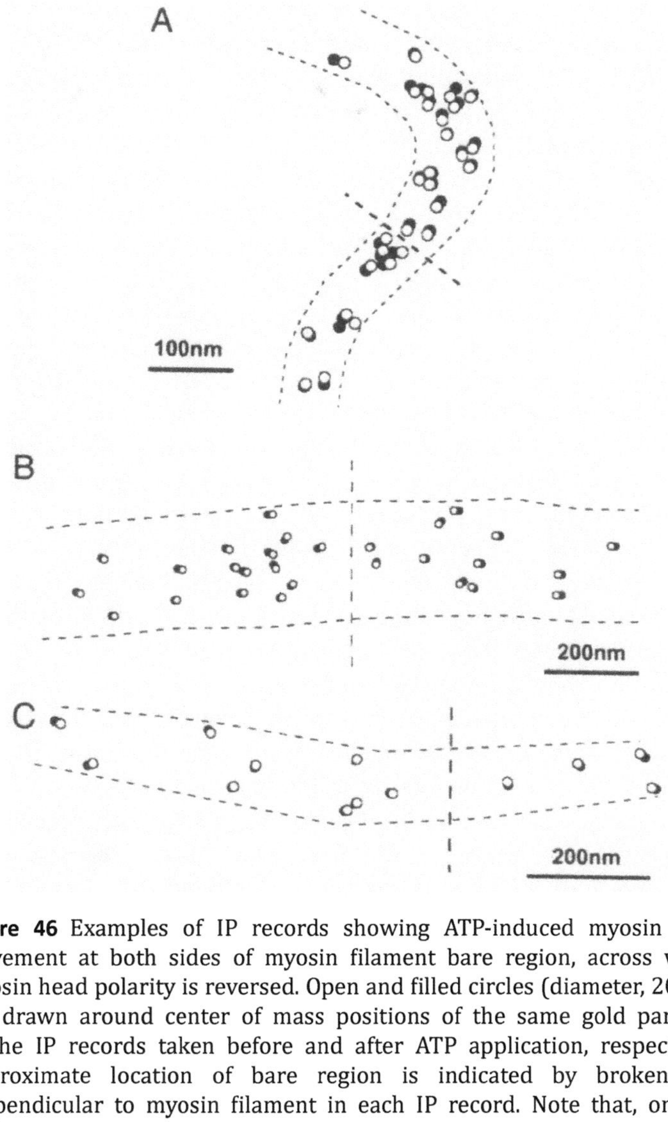

Figure 46 Examples of IP records showing ATP-induced myosin head movement at both sides of myosin filament bare region, across which myosin head polarity is reversed. Open and filled circles (diameter, 20 nm) are drawn around center of mass positions of the same gold particles in the IP records taken before and after ATP application, respectively. Approximate location of bare region is indicated by broken line perpendicular to myosin filament in each IP record. Note that, on ATP application, individual myosin heads move away from bare region. Myosin filament is thin and bent in A, while they are thick and straight in B and C. From Sugi et al. (2008).

6.3.5 Reversibility of ATP-Induced Myosin Head Movement

To ascertain whether individual myosin heads return to their initial equilibrium position after performing recovery stroke, we took three IP records of the same filament (1) before ATP application, (2) after ATP application, and (3) after complete exhaustion of ATP. Exhaustion of ATP was facilitated by adding hexokinase and D-glucose as ATP scavenger (Oiwa et al., 1971). Figure 47 shows examples of IP records, in which changes of myosin head position are shown as the change in the position of pixels containing the center of mass position of gold particle. Pixel positions before ATP application, after ATP application, and after exhaustion of applied ATP are indicated by red, blue, and yellow, respectively.

As can be seen in each frame, myosin heads first perform recovery stroke (from red to blue) on ATP application, and after exhaustion of ATP, returns toward their initial equilibrium position. In panels A, B, and I, myosin heads return exactly to their initial position, while in other frames, the return of myosin heads to their initial position is more or less incomplete, probably due to slight distortion of myosin filament structure caused by myosin head movement. These results indicate that, after the end of recovery stroke, associated with ATP hydrolysis, M-ADP-Pi $\rightarrow$ M + ADP + Pi, myosin heads tend to return to their initial position before ATP application. The return of myosin heads to their initial position can be accounted for by assuming that (1) myosin heads detached from actin filaments fluctuate around their equilibrium position, in such a way that their time-averaged mean position does not change appreciably with time, (2) on binding with ATP, they exert recovery stroke coupled with ATP hydrolysis, and (3) after the end of recovery stroke, myosin heads return to their equilibrium position. The presence of definite equilibrium position in each myosin head has been predicted in the Huxley contraction model (see Fig. 7A), as already mentioned.

When we started EC experiments using synthetic myosin filaments consisting of vertebrate skeletal muscle myosin, I retired from Teikyo University and lost my laboratory for political reasons. However, Dr. Terukazu Etoh, President of Japan Electron Optics Laboratory, Inc (JEOL) and also a pupil of Dr. Akira Fukami, very generously provided me experimental space at the factory of JEOL. In addition, I found my new coworkers Dr. Hiroyoshi Minoda in Tokyo Institute of Agriculture and Technology and Dr. Masaru Tanokura in the University of Tokyo. I cannot thank them enough for their kind considerations.

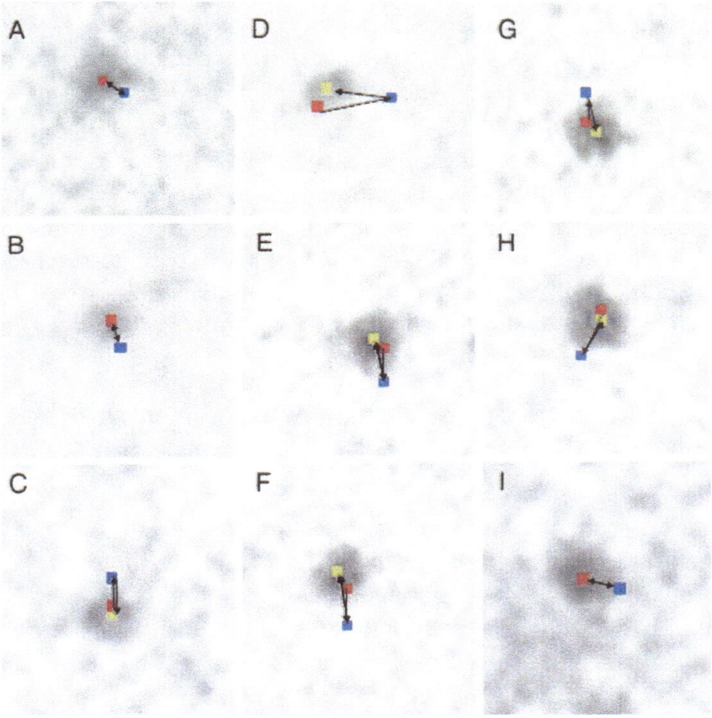

Figure 47 Changes in position of nine different pixels (each 2.5 × 2.5 nm) where the center of mass positions of gold particles are located. Pixel positions were recorded before ATP application (red), after ATP application (blue), and after exhaustion of ATP (yellow). Direction of myosin head movement is indicated by arrows. From Sugi et al. (2008).

At that time, Professor H. E. Huxley was very anxious about my situation after my retirement, but was pleased with my rather miraculous luckiness to be able to resume EC experiments at the factory of JEOL. When we succeeded in recording the myosin head recovery stroke, he was extremely happy with our success, because it proved his idea of myosin head conformational changes (Fig. 4) to be real. He sent me an e-mail, in which he stated, "*I fully appreciate that these are extraordinary difficult experiments to do, especially in your circumstances, and I fully believe that you are seeing a real phenomenon and a very interesting one. You deserve great commendation for pressing on and getting the system to work. I think the most convincing evidence is of course in your Fig. 5 in your Ms, showing clear reversal of the direction of myosin head movement on either side of what is presumably the midpoint of the myosin filament.... I really do think it is a superb piece of work and a triumph in the face of great experimental difficulties.*"

I was delighted and encouraged with the admiration of our work by Dr. Hugh Huxley, who opened the current research field on the mechanism of myofilament sliding. Furthermore, Dr. Huxley gave us a number of valuable comments when He communicated my paper on the success in recording the myosin head recovery stroke to *PNAS* (Sugi et al., 2008).

6.4 Amplitude of Myosin Head Movement at Three Different Regions of Myosin Head

The experiments with the EC hitherto described in this book are performed by using only antibody 1 attaching to the distal region of myosin head catalytic domain (CAD). To ascertain whether the myosin head COD actually keeps its angle of attachment to the actin filament, we used all three antibodies, antibodies 1, 2, and 3, attaching to (1) the distal region of myosin head CAD, (2) the reactive lysine residue (Lys 83) at the myosin head CAD, i.e., near the proximal region of myosin head CAD, and (3) the essential light chain or both essential and regulatory light chains in myosin head LD, respectively.

We succeeded in recording the ATP-induced myosin head movement at three different regions in myosin head, by attaching gold particles to myosin heads via antibodies 1, 2, or 3. The results are summarized in Fig. 48. The mean amplitude of the myosin head movement was 6.14 ± 0.09 nm (mean SEM. n = 1962) at the distal region of the myosin head CAD, 6.14 ± 0.22 nm (mean ± SEM, n = 1112) at the proximal region of myosin head CAD, and 3.55 ± 0.11 nm (mean ± SEM, n = 981), indicating that the amplitude of the movement was statistically the same between the distal and the proximal regions of the myosin head. Consequently, our results summarized in Fig. 48 were consistent with the prediction of the swinging lever arm mechanism in that myosin head CAD, attached to actin filament, moves without changing its attachment angle. Of course, this does not necessarily mean that the myosin head movement results from the active rotation of LD around COD, as predicted by the swinging lever arm hypothesis.

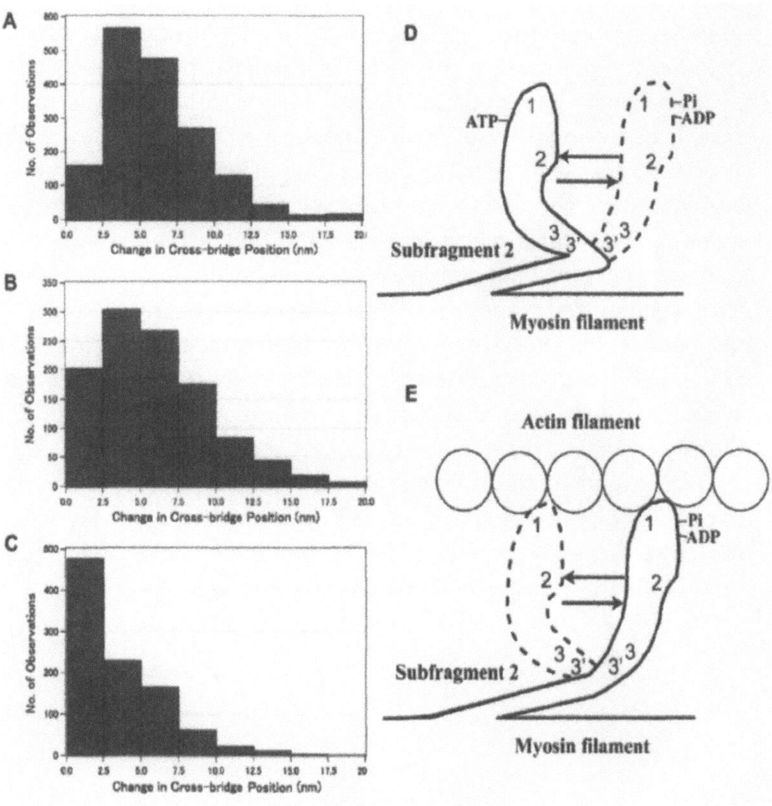

Figure 48 (A–C) Histograms showing amplitude distribution of ATP-induced movement of myosin heads, position-marked with antibody 1 (A), antibody 2 (B), and antibody 3 (C). (D and E) Diagrams showing the result that the amplitude of myosin head movement is the same between the distal and the proximal regions of myosin head, being in apparent agreement with prediction of the swinging lever arm mechanism. Approximate regions of attachment of antibodies 1, 2 and 3 and 3′ are indicated by numbers in the diagrams. From Minoda et al. (2011).

When we prepared a paper on the above results and submitted it to some mainstream journals, in which I had published many papers in the past, I was surprised that my paper was instantly rejected without reviewing procedures. Professor H. E. Huxley was also very much surprised at the instant rejection of my paper and personally informed me of an opinion of a muscle crystallographist about our EC experiments that "*Only Sugi's research group can perform such experiments with the EC, and we have to believe whatever he reports. Therefore, his papers should not be published.*" Finally, however, my paper was eventually accepted for publication in a scientific journal by the decision of the editor, who thought that the reviewers' behavior was unfair. Here, I reserve my further feelings about my bitter experience. The current appearance of many access-free scientific journals seems largely reflect the unfair behavior of reviewers and editors in mainstream journals against papers submitted by their competitors. In this connection, I am happy to use the access-free journals, which review submitted papers in a fair manner. Recently, I published my papers in such access-free journals, including *PloS ONE*, which now publishes numerous quality papers per year, obviously reflecting numerous investigators whose papers are unduly rejected by mainstream journals.

At the end of this chapter, I would like to comment that the agreement of the results in Fig. 47 with the swinging lever arm mechanism is only accidental; I do not believe this mechanism for a number of reasons described in this volume.

Chapter 7

Evidence against Current Dogmas and Mechanism of Muscle Contraction as Revealed by the Effect of Antibodies on Myosin Head

This chapter deals with the evidence against the current dogmas on the mechanism of muscle contraction. Before proceeding to my goal to record myosin head power stroke in the presence of actin filaments, it was necessary to examine the effects of the three antibodies used in the EC experiments on both Ca^{2+}-activated contraction of skinned muscle fibers and in vitro actin-myosin sliding; if the antibodies exhibit inhibitory action on muscle fiber contraction, we cannot use them for experiments with the EC to record myosin head power stroke producing muscle contraction. We found that antibodies 1 and 2 had no effect on muscle contraction, while antibody 3 inhibited muscle contraction. Unexpectedly, we also found that in vitro actin–myosin sliding was inhibited by antibody 2, but not by antibodies 1 and 3.

7.1 Antibody 1 Has No Effect on Both Muscle Contraction and in vitro Actin–Myosin Sliding

As already explained, antibody 1 binds with the junctional peptide between 50K and 20K fragments of myosin heavy chain, which is

Mysteries in Muscle Contraction: Evidence against Current Dogmas
Haruo Sugi

ISBN 978-981-4774-48-2 (Hardcover), 978-1-315-10191-0 (eBook)
www.panstanford.com

located in the vicinity of two main actin-binding sites in myosin head CAD, as shown in Fig. 15. Consequently, if the bulky antibody 1 (IgG) molecule is attached to the junctional peptide, the two main actin-binding sites are completely covered with the IgG molecule to inhibit the formation of actin–myosin rigor complex, A-M, which is believed to exist in the Lymn–Taylor scheme (Fig. 18) following the release of ADP and Pi from A-M-AP-Pi complex, as A-M-ADP-Pi → A-M + ADP + Pi.

We first examined the effect of antibody 1 on Ca^{2+}-activated contraction of skinned skeletal muscle fibers. Unexpectedly, antibody 1 had no appreciable effect on both the maximum unloaded shortening velocity and the maximum isometric force of Ca^{2+}-activated muscle fibers even at high concentrations (up to 2 mg/ml) (Fig. 49).

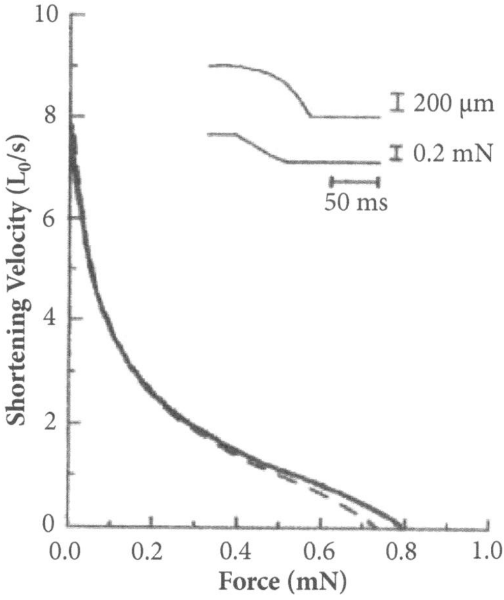

Figure 49 No appreciable effect of antibody 1 on the relation between shortening velocity (*V*) and force (*P*) in Ca^{2+}-activated muscle fibers. Inset shows method of determining the force–velocity relation by applying ramp decrease in force to the fiber. Solid and broken lines indicate *P*–*V* relations in the absence and in the presence of antibody (2 mg/ml), respectively. From Sugi et al. (2014).

We also observed no appreciable effect of antibody 1 on ATP-dependent sliding velocity of fluorescently labeled actin filaments on myosin molecules fixed on a glass surface, and found that antibody 1 at concentrations up to 2.0 mg/ml had no significant effect on the in vitro actin–myosin sliding, as can be seen in Fig. 50. These unexpected findings made me skeptical about the widely believed dogma that during cyclic attachment and detachment between myosin heads and actin filaments (Lymn–Taylor scheme, Fig. 18), myosin heads (M) pass through the rigor configuration (A-M) during the reaction, A-M-ADP → AM + ADP. Of course, the complex A-M in this reaction does not necessarily take a configuration similar to that constructed from extracted protein samples.

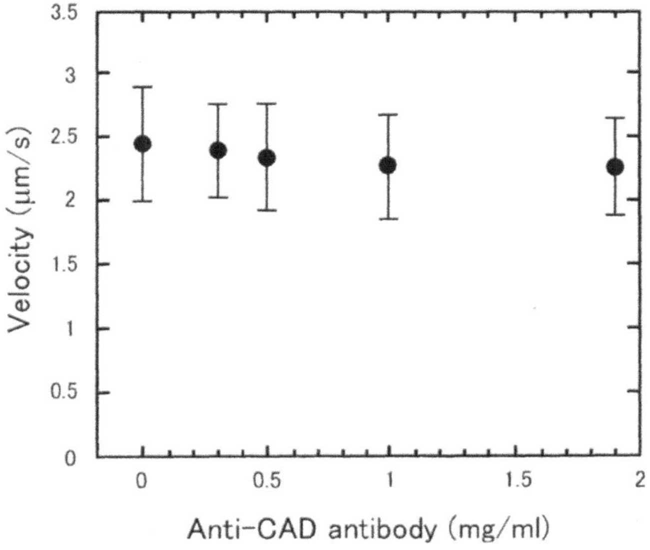

Figure 50 No appreciable effect of antibody 1 (anti-CAD antibody) on the ATP-dependent sliding of actin filaments on myosin fixed on a glass surface. Mean velocities of actin filament sliding with SD (n = 80–120) are plotted against antibody concentration. From Sugi et al. (2014).

7.2 Antibody 2 Inhibits in vitro Actin–Myosin Sliding but Has No Effect on Muscle Contraction

In contrast with the lack of appreciable effect of antibody 1 on both muscle contraction and in vitro actin–myosin sliding, antibody 2 was found to have marked inhibitory effect on in vitro actin–myosin sliding, while it had no appreciable effect on muscle contraction.

As shown in Fig. 51, ATP-dependent in vitro sliding of actin filaments over myosin molecules fixed on a glass surface was completely inhibited by antibody 2, attaching to reactive lysine residue (RLR) in myosin head COD, at a concentration as low as 0.15 mg/ml. Meanwhile, the actin-activated myosin head (S-1) ATPase activity remained unchanged in the presence of antibody 2 (0.36 mg/ml) (Fig. 52), indicating that myosin head (S-1) still exhibit actin-activated ATPase activity in the presence of antibody 2. This result indicates dissociation of ATPase activity and actin–myosin sliding; although myosin head CAD can still hydrolyze ATP at a rate similar to that in the absence of antibody 2, the resulting myosin head power stroke producing sliding of actin filaments over myosin molecules disappears in the presence of antibody 2.

On the other hand, antibody 2 had no appreciable effect on Ca^{2+}-activated muscle fiber contraction as with antibody 1 (Fig. 53). These findings indicate that the mechanism of in vitro ATP-dependent actin–myosin sliding is completely different from that of muscle contraction, which results from the attachment–detachment cycle between myosin heads and actin filaments in the 3D myofilament lattice. In muscle, each myosin head is surrounded by six actin filaments in close apposition to it, while the sequence of events taking place in ATP-dependent actin filament sliding on myosin, HMM, or S-1 fixed on a glass surface is not clear (H. E. Huxley, 1990).

As the mechanism of zigzag actin filament sliding over myosin molecules randomly fixed on the substratum is unclear (H. E. Huxley, 1990), the marked difference in the effect of antibody 2 between muscle contraction and in vitro actin–myosin sliding can be taken to indicate that the performance of randomly fixed myosin heads in producing actin filament sliding is completely different from that of myosin heads in the 3D myofilament lattice structure in the muscle. It seems likely that myosin heads fixed on the substratum are much more flexible than those in the myofilament lattice, so that they can markedly change their configuration to interact the with actin filament passing through them in random directions; such flexibility of randomly fixed myosin heads would easily be lost by antibody 2 attaching to the reactive lysine residue in their converter domain.

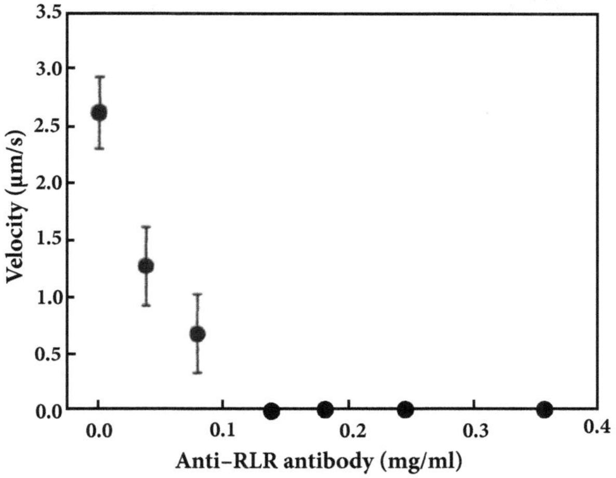

Figure 51 Marked inhibitory effect of antibody 2 attaching to reactive lysine residue (RLR) in myosin head COD (anti-RLR antibody) on in vitro actin–myosin sliding. From Sugi et al. (2014).

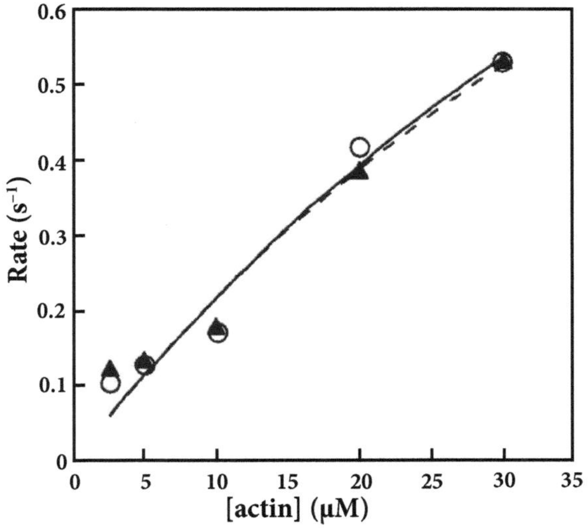

Figure 52 No appreciable effect of antibody 2 on the actin-activated myosin head (S-1) ATPase activity. Open circles and filled triangles show rate of ATP hydrolysis of actin-activated S-1 ATPase in the absence and in the presence of antibody 2 (0.36 mg/ml), respectively. From Sugi et al. (2014).

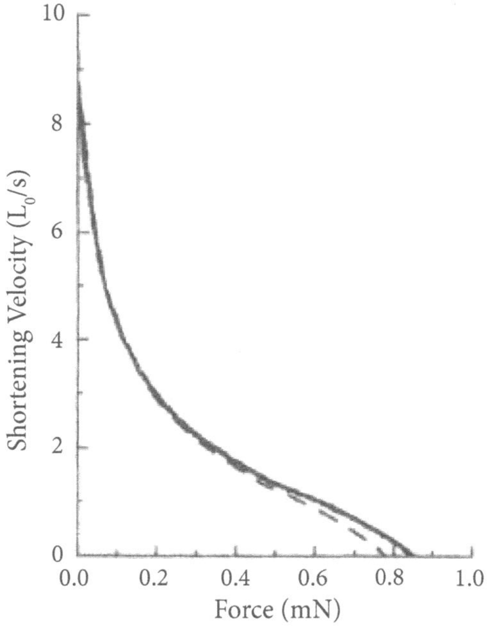

Figure 53 No appreciable effect of antibody 2 on the *P*–*V* relation of Ca^{2+}-activated muscle fibers. Solid and broken lines indicate *P*–*V* relations in the absence and in the presence of antibody 2 (2 mg/ml). From Sugi et al. (2014).

7.3 Antibody 3 Has No Significant Effect on in vitro Actin–Myosin Sliding but Inhibits Isometric Force Development in Muscle Without Changing ATPase Activity

As shown in Fig. 54, antibody 3 (anti-LD antibody), attaching to two light chains in myosin head LD, showed no significant effect on ATP-dependent in vitro actin–myosin sliding even in high concentrations (up to 2 mg/ml). Although the velocity of actin filament sliding on myosin exhibited a tendency to decrease with increasing antibody concentration, the difference between the data points was not statistically significant.

On the other hand, antibody 3 showed inhibitory effect on Ca^{2+}-activated isometric force development of muscle fibers in a concentration-dependent manner. As shown in Fig. 41,

Ca^{2+}-activated isometric force was reduced markedly by up to 80% in the presence of 2 mg/ml antibody. The inhibitory effect of antibody 3 was reversible; when the fibers were returned to relaxing solution without antibody, and kept in it for 20 min, they completely restored their ability to develop isometric force as large as that before application of the antibody. This indicates that the antibodies attached to myosin heads can be readily be removed. This point is important when the reversible antibody effects are compared with that of irreversible chemical modifications, which may alter structures in the vicinity of the regions of chemical modification.

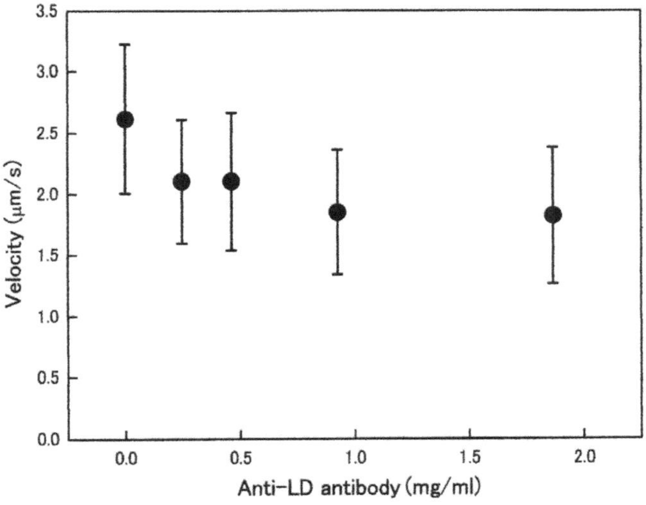

Figure 54 No significant effect of antibody 3 (anti-LD antibody) attaching to two light chains in myosin head LD. From Sugi et al. (2014).

Figure 56A shows the effect of antibody 3 on the force–velocity relation of Ca^{2+}-activated muscle fibers. Despite the marked reduction of the maximum isometric force in the presence of antibody 3 (2 mg/ml), the maximum velocity of shortening, V_{max}, remained unchanged. If the force–velocity curves were normalized with respect to the maximum isometric force, the two force–velocity curves were found to be identical, as can be seen in Fig. 56B. This indicates that the reduction of isometric force by antibody 3 results from reduction in the number of myosin heads involved in isometric force generation, implying that myosin heads with antibody 3 attached can no longer generate the isometric force.

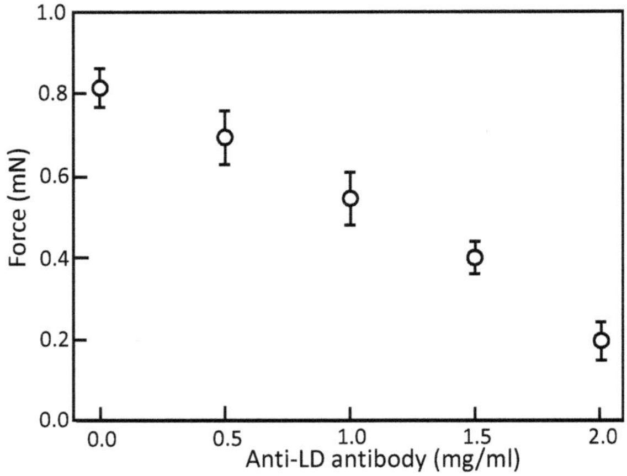

Figure 55 Inhibitory effect of antibody 3 (anti-LD antibody), attaching to two light chains in myosin head LD on Ca^{2+}-activated isometric force development of muscle fibers. Amplitude of isometric force is plotted against antibody concentration. From Sugi et al. (2014).

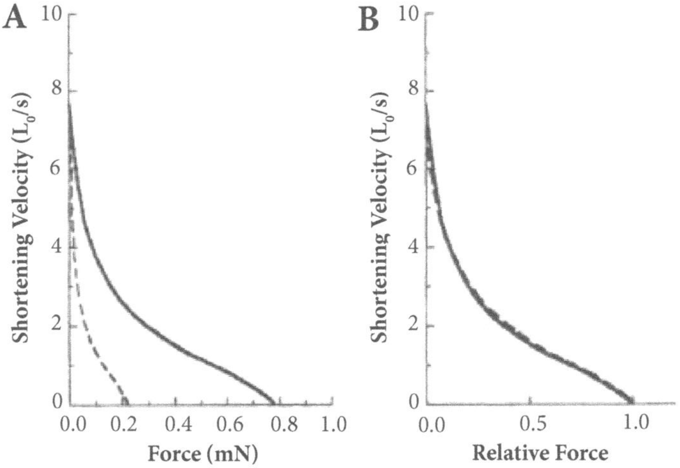

Figure 56 Effect of antibody 3 on force–velocity relation of Ca^{2+}-activated muscle fibers. (A) Force–velocity curves before (solid line) and after (broken line) application of antibody 3 (1.5 mg/ml). (B) The same force–velocity curves as in A normalized with respect to the maximum isometric force. Note complete overlap of two curves. From Sugi et al. (2014).

We also examined the effect of antibody 3 on the MgATPase activity of Ca^{2+}-activated muscle fibers by recording MgATPase activity by decrease of NADH fluorescence during ATP hydrolysis (Sugi et al., 1992). Figure 57 shows simultaneous recordings of MgATPase activity (upper traces) and Ca^{2+}-activated isometric force development (lower traces) of muscle fibers. It can be seen that despite the marked reduction of isometric force, MgATPase activity of muscle fibers, as measured from the slope of ATPase records, does not change significantly in the presence of antibody 3 (2 g/ml).

This result is explained to be due to myosin heads, with antibody 3 attached, still hydrolyzing ATP at a rate similar to that of myosin heads generating normal isometric force. The dissociation of contractile activity and ATPase activity in myosin heads by antibody 3 may result from that the ATP-binding site in myosin head CAD is distant from myosin head LD, to which antibody 3 attaches.

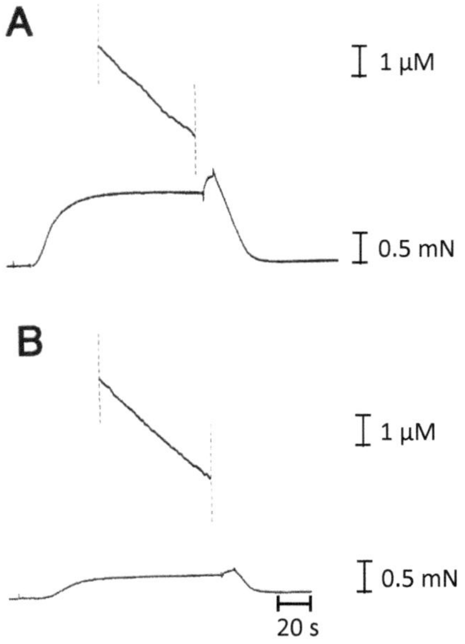

Figure 57 Simultaneous recordings of MgATPase activity (upper traces) and Ca^{2+}-activated isometric force (lower traces) before (A) and after (B) application of antibody 3 (3 mg/ml). Note that MgATPase activity does not change appreciably despite marked reduction of isometric force. From Sugi et al. (2014).

7.4 Interpretation of the Effect of Antibodies on Muscle Contraction and in vitro Actin–Myosin Sliding

In this section, I will consider the different effects of antibodies 1, 2, and 3 on muscle contraction and in vitro actin–myosin sliding, to explain how the results obtained constitute evidence against current dogmas on the mechanism of muscle contraction. Before starting discussions, it seems necessary to present evidence that antibody (IgG) molecules can actually diffuse into skinned muscle fibers to bind with their respective epitopes.

7.4.1 Diffusion of Antibody (IgG) Molecules into Skinned Muscle Fibers

The possibility that antibody molecules do not diffuse into skinned muscle fibers is ruled out by the following reasons: (1) using the confocal microscope, fluorescently labeled antibody molecules (IgG) are shown to diffuse into skinned muscle fibers, and evenly distribute in the interior of skinned fiber within a few minutes (Kraft et al., 1995); (2) antibody 3 (IgG), having the same structure as antibodies 1 and 2 except for the epitope binding site, shows the reversible inhibitory effect on muscle contraction (Figs. 55–57), indicating that it actually diffuses into and out of skinned fibers; and (3) when relaxed muscle fibers are put into rigor solutions containing antibodies 1, 2, or 3, much more time is required to establish the rigor state compared to the establishment of the rigor state in rigor solution without antibodies. The above effect of antibodies in slowing establishment of rigor state may result from that antibodies diffuse into the fibers and bind with their respective epitopes on myosin heads until they are overridden by rigor A-M linkages (Sugi et al., unpublished results). The fairly high affinity of antibodies to their epitopes in myosin heads is indicated by comparing the number of myosin heads position marked with antibodies, and the estimated number of myosin heads in synthetic myosin filaments (Sugi et al., 2015).

In the case of in vitro actin–myosin sliding, on the other hand, there are no barriers against diffusion of antibodies to access myosin heads fixed on a glass surface.

7.4.2 Ineffectiveness of Antibody 1 on Both Muscle Contraction and in vitro Actin–Myosin Sliding

In our experiments described in this chapter, antibody 1 showed no appreciable effect on both muscle contraction and in vitro actin–myosin sliding (Figs. 49 and 50). Since antibody 1 attaches to the junctional peptide between 50K and 20K fragments of myosin heavy chain in myosin head CAD, and covers two main actin-binding sites on myosin (Fig. 20), its ineffectiveness in inhibiting both muscle contraction and in vitro actin–myosin sliding indicates that during muscle contraction, myosin heads do not pass through the rigor A-M configuration corresponding to the structure of the acto-S-1 complex prepared from extracted protein samples (Fig. 19).

The prolongation of time of establishment of rigor state in skinned muscle fibers supports the actual attachment of antibody 1 to cover actin-binding sites in myosin head CAD; though the antibody molecules attached to myosin heads are finally overridden by rigor A-M linkage formation, this process takes place on the time scale of many seconds. It seems therefore likely that during muscle contraction, myosin heads take a structure different from that of the acto-S-1 complex after the end of the power stroke, as already suggested by Radocaj et al. (2009).

To obtain information about this issue, Sugi and coworkers compared mechanical response to applied fiber length changes between muscle fibers put into rigor state by removing ATP from contracting solution (high-Ca rigor) and muscle fibers put into rigor state by removing ATP from relaxing solution (low-Ca rigor). It was assumed that in high-Ca rigor fibers, myosin heads were brought into rigor state after finishing their last power stroke in contracting solution, while in low-rigor fibers, myosin heads form rigor A-M linkages by overriding tropomyosin inhibiting A-M interaction in relaxing solution.

As can be seen in Fig. 58, both high-Ca and low-Ca rigor muscle fibers exhibited small but distinct force recovery following ramp-shaped releases (0.5% of fiber length, complete in 5 ms), which resembled in time course the quick force recovery following quick release in intact contracting muscle fibers (Fig. 8), though the time scale was about three orders of magnitude slower in the former than in the latter.

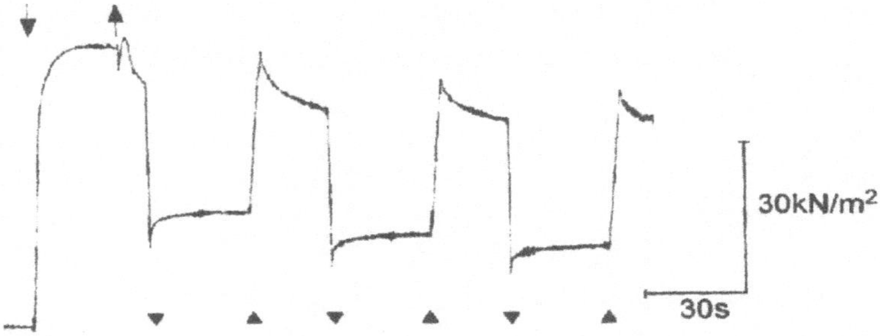

Figure 58 Force recovery following ramp-shaped releases (amplitude, 0.5% of fiber length; duration, 5 ms) in single skinned muscle fibers in high-Ca rigor state. The fiber was first made to contract isometrically in contracting solution (downward arrow) and then put into rigor solution (upward arrow) to be subjected to a series of release-stretch cycles indicated by downward and upward arrowheads. From Sugi et al. (2016).

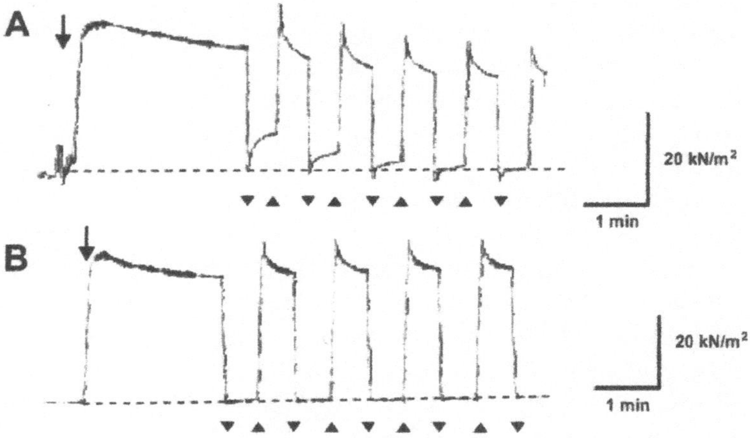

Figure 59 Force recovery in low-Ca rigor fibers. In (A) the fiber was put into low-Ca rigor state by removing external ATP (arrow), and after development of rigor force, subjected to a series of release-stretch cycles. In (B), the fiber was put into low-Ca rigor state by removing not only ATP, but also external Mg^{2+} with EDTA. Note the disappearance of force recovery following release. From Sugi et al. (2016).

The force recovery in low-Ca rigor fibers disappeared quickly by removing external Mg^{2+} with EDTA (Fig. 59), while the force recovery in High-Ca rigor fibers remained unchanged over many minutes after removal of external Mg^{2+}. These results suggest that in high-Ca rigor fibers, most myosin heads may be in the state of A-M-ADP having a long lifetime and somewhat dynamic properties, which show up as force recovery following release, while in low-Ca rigor fibers, A-M-ADP myosin heads have a short lifetime and readily disappear on removal of external Mg^{2+} with EDTA.

7.4.3 Marked Inhibitory Effect of Antibody 2 on in vitro Actin–Myosin Sliding

We found that antibody 2, attaching to reactive lysine residue (Lys 83) in myosin head converter domain, completely inhibited in vitro actin–myosin sliding (Fig. 51) without affecting actin-activated myosin head (S-1) ATPase activity (Fig. 52). The striking dissociation of actin–myosin sliding from ATP hydrolysis by antibody 2, together with its ineffectiveness in affecting muscle contraction (Fig. 53), revealed a definite difference in the mechanism between muscle contraction and in vitro actin myosin sliding.

This striking result can be accounted for as being due to the difference in condition of myosin molecules between the muscle and in vitro motility assay system; in the former, myosin heads are connected to myosin filament backbone via subfragment-2 (S-2), and facing actin filaments in 3D myofilament-lattice structure (Figs. 2 and 3), whereas myosin molecules, in the form of full-length myosin, HMM or S-1, are randomly fixed on the substratum with uncertain manner of fixation and orientation. As a result, myosin heads should be flexible to catch actin filaments, approaching toward them in a zigzag manner. The flexibility of the myosin head may be largely associated with their conformational changes at its distal region, i.e., rotational movement of myosin head CAD around COD, serving as the pivot. If this explanation is correct, the inhibitory effect of antibody 2 on in vitro actin–myosin sliding can be understood as resulting from the attachment of antibody 2 to the reactive lysine residue in myosin head COD to inhibit rotation of CAD around COD.

In contrast, myosin heads in the muscle are in close vicinity of actin filaments in the regularly arranged 3D myofilament-lattice, and it is not necessary for them to change their conformation markedly by rotating CAD around COD to interact with actin filaments. Consequently, the attachment of antibody 2 to the reactive lysine residue in COD does not affect muscle contraction.

This implies that the mechanism of in vitro actin–myosin sliding differs too far from that in muscle contraction, and the results obtained from in vitro actin-myosin sliding may have no direct relation with what is actually taking place in contracting muscle. More or less the same argument would apply to the optical trap method because of uncertainties about fixation and orientation of myosin heads.

Muhlrad et al. (2003) reported that chemical modification (trinitrophenylation) of lysine residue in myosin head COD inhibited in vitro actin–myosin sliding. According to them, this result was due to mechanical collision between structures in COD. They reported, however, that actin-activated ATPase activity of myosin head CAD was also inhibited by the chemical modification of lysine residue, in contrast with our result that antibody 2 had no appreciable effect on actin-activated S-1 ATPase activity. The difference strongly suggests that chemical modification of lysine residue may change the 3D structure of myosin head to inhibit ATPase activity of CAD, while the reversible mild binding of antibody 2 to lysine residue does not produce any myosin head structural changes. Thus, our results cast doubt on the essential role of myosin head COD in muscle contraction.

7.4.4 Inhibitory Effect of Antibody 3 and Anti-S-2 Antibody on Muscle Contraction

As already described, the inhibitory effect of antibody 3 on muscle fiber force generation (Figs. 55 and 56) without changing the MgATPase activity (Fig. 57) indicates that, when antibody 3 molecules attach to myosin head LD, the movement at myosin head LD-S-2 boundary may be inhibited, so that myosin heads can no longer participate in isometric force generation, while myosin head CAD keeps its binding with actin filaments to exhibit normal rate of actin-activated ATPase activity.

Then, two questions arise: (1) How can the myosin heads with antibodies attached maintain their high ATPase activity without tightly attaching to actin filaments? (2) How can myosin head LD or myosin S-2 transmit information to the actin-binding site of myosin head CAD? Concerning the second question, we have already carried out experiments, showing the existence of functional communication between myosin head CAD and myosin S-2 via myosin head LD, as described later.

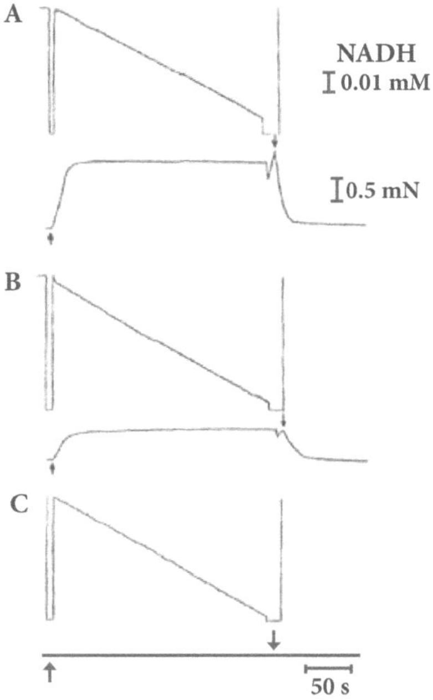

Figure 60 Simultaneous recordings of MgATPase activity (upper traces) and Ca^{2+}-activated isometric force (lower traces) of skinned muscle fibers before (A) and at 100 (B) and 150 min (C) after application of anti-S-2 antibody (1.5 mg/ml). Even when isometric is reduced to zero, MgATPase activity, measured by decrease of NADH during ATP hydrolysis, remains unchanged. From Sugi et al. (1992).

In 1992, I was asked by Professor Harrington, who published many papers on the possible role of myosin-S-2 over many years, to work together on the effect of polyclonal antibody to myosin S-2 (anti-S-2 antibody) prepared in his laboratory. First, we tested the effect of anti-S-2 antibody on in vitro actin–myosin sliding, and found no appreciable effect. Next, we examined its effect on Ca^{2+}-activated contraction in skinned muscle fibers with the following interesting results: (1) In the presence of the anti-S-2 antibody, the magnitude of Ca^{2+}-activated isometric force decreased gradually with time to eventually disappear completely

(Fig. 60). (2) In contrast, the anti-S-2 antibody showed no appreciable effect on the MgATPase activity of Ca^{2+}-activated muscle fibers (Fig. 60). (3) The force–velocity curves obtained at different magnitudes of isometric force were found to be identical when normalized with respect to the maximum force levels (Fig. 61) (Sugi et al., 1992).

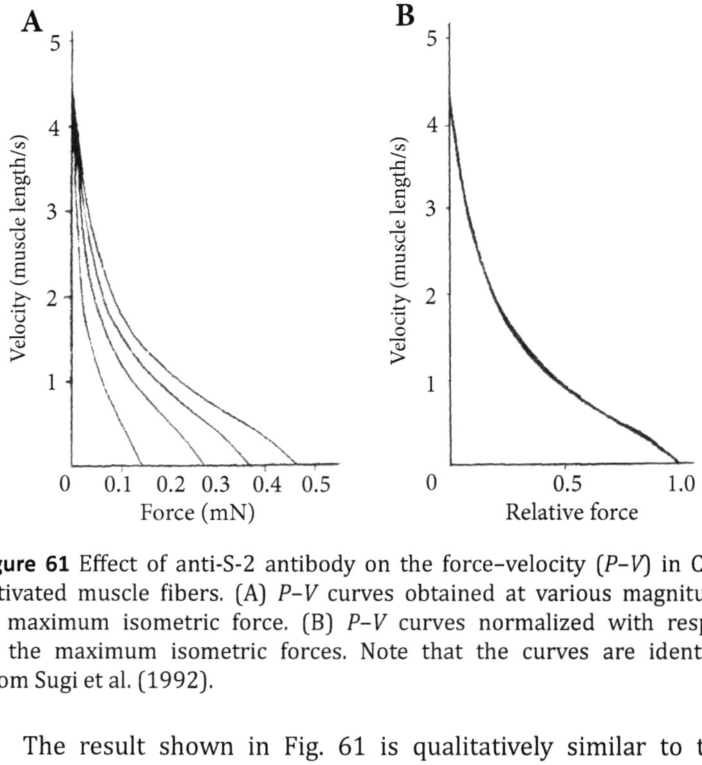

Figure 61 Effect of anti-S-2 antibody on the force–velocity (*P–V*) in Ca^{2+}-activated muscle fibers. (A) *P–V* curves obtained at various magnitudes of maximum isometric force. (B) *P–V* curves normalized with respect to the maximum isometric forces. Note that the curves are identical. From Sugi et al. (1992).

The result shown in Fig. 61 is qualitatively similar to that shown in Fig. 57. This indicates that irrespective of whether antibodies attach to myosin head LD or myosin S-2, located at both sides of the LD-S-2 boundary, myosin heads, with either antibody attached, become unable to generate isometric force.

7.4.5 Evidence for Functional Communication between Myosin Head CAD and Myosin S-2 Regions

To ascertain whether functional communication can actually be transmitted from the S-2 and the LD regions to actin-binding site in the CAD region within a myosin head, we made experiments, in which anti-S-2 antibody was applied to skinned fibers after development of Ca^{2+}-activated force or after the development of rigor force after removal of ATP from relaxing solution (Fig. 62) (Kobayashi et al., 1998).

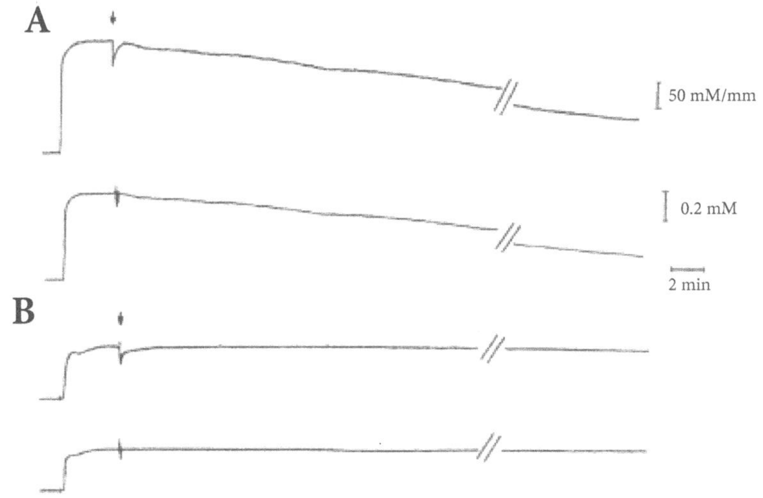

Figure 62 Stiffness (upper traces) and isometric force (lower traces) records of muscle fibers when anti-S-2 antibody is applied after development of steady Ca^{2+}-activated force (A) and after development of rigor force (B). Times of application of antibody is indicated by downward arrows. From Kobayashi et al. (1998).

After antibody application, both muscle fiber stiffness and Ca^{2+}-activated isometric force started decreasing toward zero with time (Fig. 62A), being consistent with the idea that attachment of antibody to myosin S-2 region in individual myosin heads influence actin-binding site in myosin head CAD region to attach to actin filament.

If, on the other hand, the antibody is applied after the development of rigor force, it had no appreciable inhibitory effect on both muscle fiber stiffness and rigor force (Fig. 62B); both stiffness and force maintained their steady level in the presence of antibody. This indicates that if strong actin–myosin binding is already established at the myosin head CAD region, the attachment of antibody to myosin S-2 region participating in rigor force production cannot influence myosin head CAD to change its mode of binding with actin filament from strong to weak state. At present, the functional communication between myosin head CAD and myosin S-2 is a mystery to be studied in future.

Meanwhile, the absence of significant effect of antibody 3 on in vitro actin–myosin sliding (Fig. 54) may result simply because the myosin S-2 region is immobilized in most myosin molecules fixed on the substratum; the tendency of actin–myosin sliding velocity to decrease with increasing antibody 3 concentration might result because in some myosin molecules on the substratum, the myosin S-2 region can move and participate in actin–myosin sliding.

7.4.6 Regulation of Myosin Head-Actin Binding Strength by Myosin Head LD and Myosin S-2

To obtain further evidence for the functional communication between myosin S-2 and myosin head CAD regions, we performed experiments using the actin cable versus the myosin system, combined with the centrifuge microscope (Fig. 22). We measured sliding velocities of myosin-coated polystyrene beads (diameter, 2.8 or 4.5 μm), previously treated with anti-S-2 antibody (Tsuchiya et al., 1998). Figure 63 shows the selected frames from a video film showing the ATP-dependent movement of beads coated with anti-S-2 antibody-treated myosin or untreated (control) myosin. It can be seen that the bead coated with untreated myosin moves faster than the bead coated with myosin previously treated by anti-S-2 antibody (1.0 mg/ml). Under zero centrifugal force, the velocity of myosin-coated beads decreased with increasing antibody concentration, with which myosin was previously treated, and finally stopped moving when the antibody concentration exceeded 1.5 mg/ml (Fig. 64).

Using the centrifuge microscope system, it was possible to measure the forces required to detach myosin-coated beads from actin cables, namely to measure and compare the strength of binding of myosin heads to actin cables. In the beads coated with myosin, previously treated by anti-S-2 antibody > 1.5 mg/ml, did not move in the presence of 2 mM ATP and remained attached to actin cables (Fig. 64). By applying centrifugal forces in the direction opposite to the bead movement to serve an external load against the bead movement, the force (=load) velocity curves were obtained. As shown in Fig. 65, beads coated with antibody-treated myosin could be stopped and detached from actin cables with very small forces from 0.3 to 3.5 pN (1.1 ± 0.8 pN, mean ± SD, n = 7), compared to the forces required to stop moving beads coated with untreated myosin, i.e., the maximum "isometric" force generated by control myosin molecules, ranging from 8.5 to 28 pN (10.2 ± 4.8 pN, mean ± SD, n = 8).

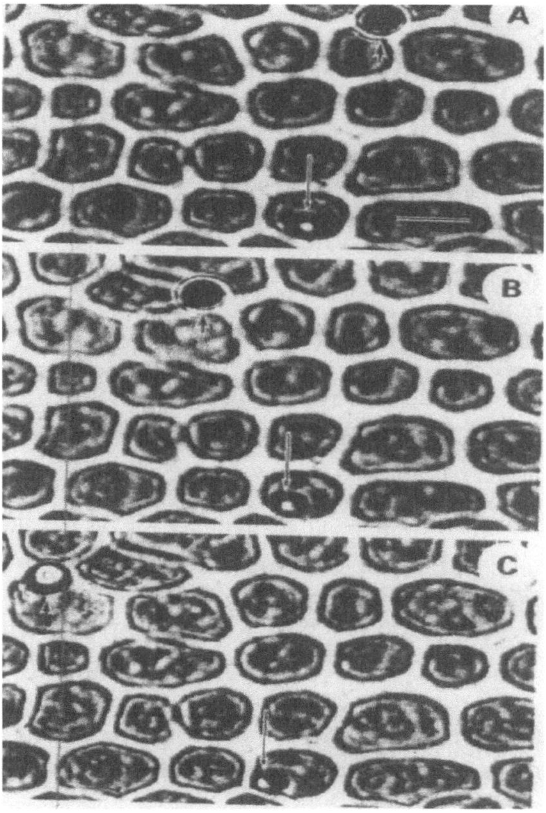

Figure 63 Selected video frames showing two myosin-coated polystyrene beads sliding from right to left along chloroplast rows on which actin cables run straight. One bead (diameter, 4.5 μm, indicated by short upward arrows) was coated with control myosin, while the other bead (diameter, 2.8 μm, indicated by long downward arrows) was coated with myosin previously treated by 1 mg/ml anti-S-2 antibody. The interval between the frames is 4 s. Note that the bead coated with control myosin moves much faster than the bead coated with myosin previously treated by 1.0 mg/ml anti-S2 antibody. From Tsuchiya et al. (1998).

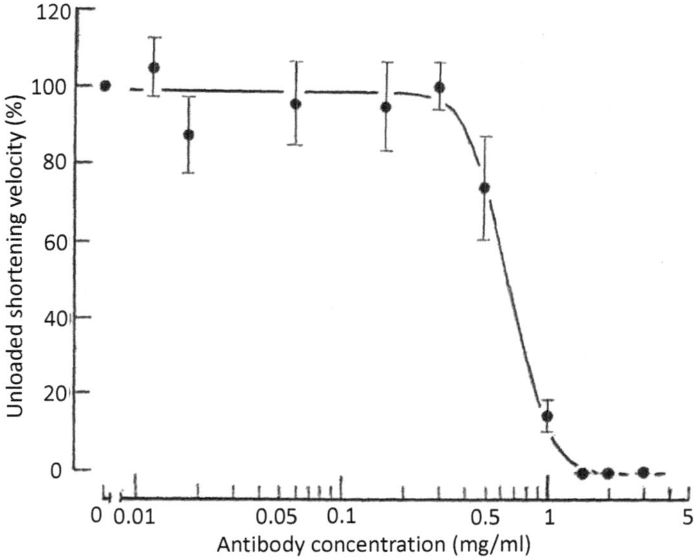

Figure 64 Effect of anti-S-2 antibody on ATP-dependent steady unloaded velocity of polystyrene beads coated with myosin, previously treated with antibody in various concentrations. Velocities are expressed relative to control value. Each data points represents mean ± SD (n = 8~15). From Tsuchiya et al. (1998).

If beads coated with untreated myosin are placed on actin cables in the absence of ATP, they remain attached to actin cables by forming rigor actin–myosin linkages. Such beads could not be detached even with the maximum force of 40 pN generated by the centrifuge microscope. These results demonstrate that the anti-S-2 antibody attached to the myosin S-2 region can exert influence on the actin-binding sites in myosin head CAD, to keep weak binding with actin filaments. The specific attachment of anti-S-2 antibody along the S-2 region of myosin molecule has been confirmed by rotary shadowed myosin molecules (Katayama, unpublished observation).

In summary, it is clear that myosin S-2 plays an essential role in muscle contraction, and it therefore seems strange why crystallographists, who constructed the swinging lever arm mechanism, ignore the important function of myosin S-2 region in regulating strength of myosin CAD binding to actin without any definite reasons.

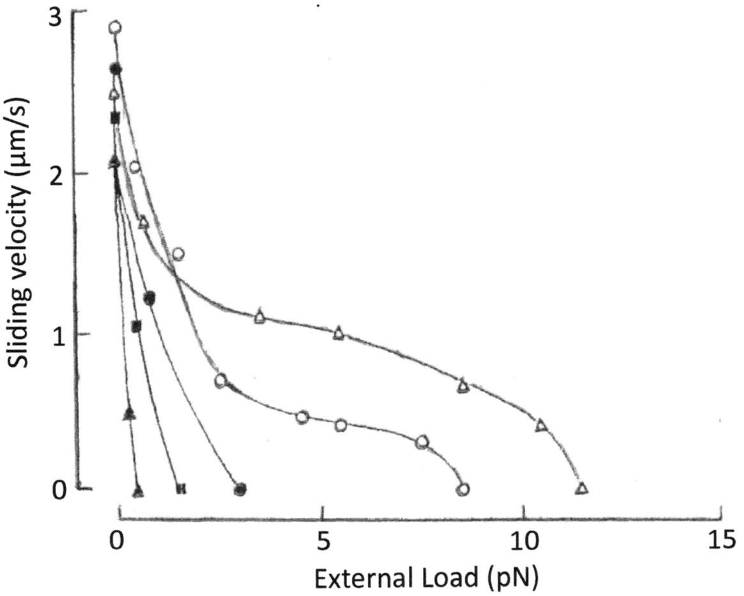

Figure 65 Steady-state relation between ATP-dependent sliding velocity of myosin-coated beads and the external load. Open symbols represent data points with control untreated myosin. Filled symbols represent data points with myosin treated with antibody at 0.2 mg/ml (filled circles), 0.3 mg/ml (filled squares), and 0.5 mg/l (filled triangles), respectively. From Tsuchiya et al. (1998).

7.4.7 Evidence for Cooperativity of Myosin Two Heads in Force Generation in Muscle

In this section, I will explain our experiments on the possible cooperative action of myosin two heads in force generation in muscle, because the results obtained are closely related to the effect of antibody 3 and anti-S-2 antibody on the strength of myosin CAD binding to actin filament.

It still remains to be a mystery why myosin molecules in the muscle have two S-1 heads (Fig. 4). In the case of porter myosins, which convey substances along actin filaments in the interior of cells in the form of single molecules, their processive movement along actin filaments is known to be like a person walking with two legs with a constant stride amplitude without detaching from the actin filament (Fig. 66).

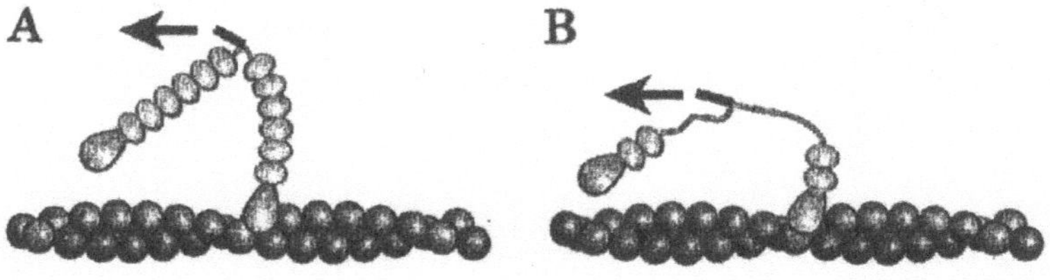

Figure 66 Processive movement of two kinds of porter myosin molecules along actin filaments. From Sugi and Kobayashi (2006).

On the other hand, myosin heads extending from myosin filaments perform power strokes asynchronously and are compared to rowers in a boat with asynchronous rowing motion. This makes it difficult to investigate whether the two myosin heads of each myosin molecule work cooperatively like two heads of a porter myosin or work independently from each other. Up to the present time, most theories of muscle contraction assume individual myosin heads as independent force (and motion) generators.

To obtain information about possible cooperativity of myosin two heads, we studied the effect of p-phenylenedimaleimide (p-PDM), which is known to react with reactive sulfhydrils on myosin head to eliminate its binding to actin filament and its MgATPase activity in skinned skeletal muscle fibers (Chaen, Shimada, and Sugi, 1986). Partially p-PDM-treated muscle fibers contain a mixture of myosin molecules with zero, one, or two of their heads inactivated with p-PDM (Tokiwa and Morales, 1971; Reisler, 1980). If the two heads of each myosin molecule are assumed to react with p-PDM independently, the fraction of the three species of myosin molecules in the fibers can be estimated from simple probabilistic calculations illustrated in Fig. 67. The validity of this assumption has been verified by Reisler (1980), who showed that in partially p-PDM-treated HMM samples, the fraction of two head-modified myosin molecules was proportional to the square of the total fraction of reacted myosin heads.

The magnitude of Ca^{2+}-activated isometric force in p-PDM-treated muscle fibers, expressed relative to control value of untreated (control) fibers as T/T_0, was found to be proportional to the square of MgATPase activity, also expressed relative to the control value as J/J_0 (Fig. 68). This result can be readily explained

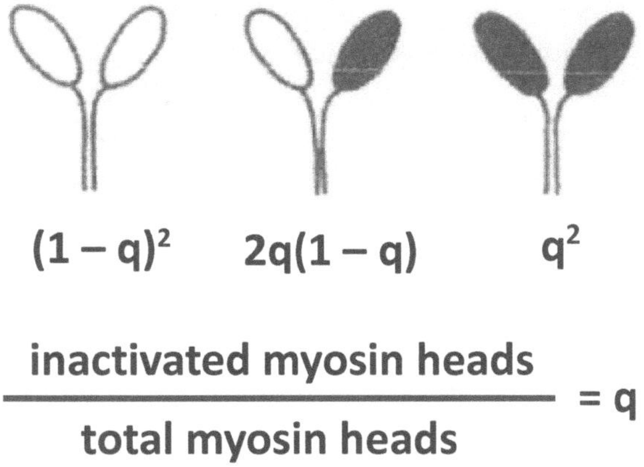

Figure 67 Three different species of myosin molecules and their respective fraction calculated on the assumption of independent reaction of myosin two heads with p-PDM. White and black myosin heads represent native and p-PDM-inactivated ones, respectively. From Chaen, Shimada, and Sugi (1986).

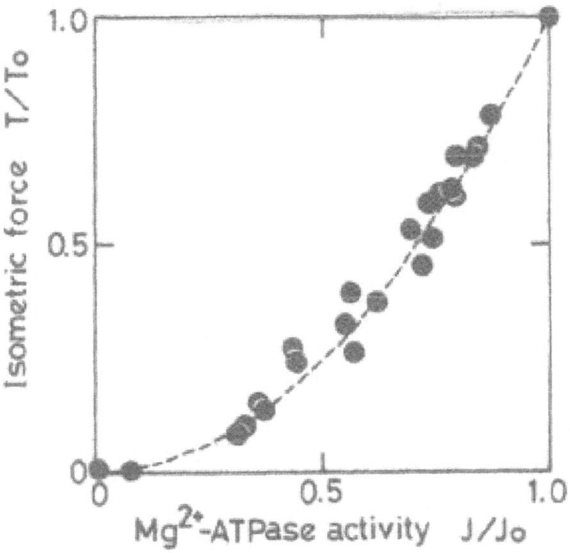

Figure 68 Relation between Ca^{2+}-activated isometric force T and MgATPase activity J in p-PDM-treated muscle fibers. Both values of T and J are expressed relative to control values T_0 and J_0, respectively. Broken line shows the curve calculated from equation, $T/T_0 = (J/J_0)^2$. From Chaen, Shimada, and Sugi (1986).

by assuming that the MgATPase activity in p-PDM-treated fibers is proportional to the total fraction of native myosin heads $(1-q)$, while the isometric force is proportional to the value $(1-q)^2$; namely, all native myosin heads hydrolyze ATP independently, while the isometric force is generated only by myosin molecules with two native heads, being consistent with the probabilistic predictions (Fig. 69).

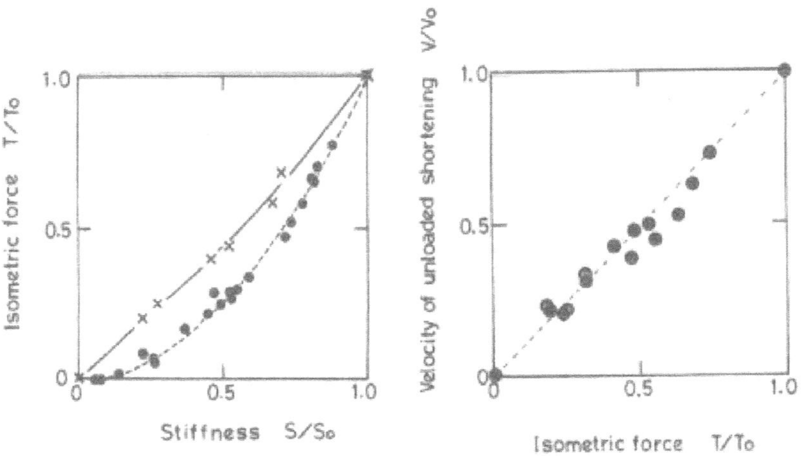

Figure 69 (Left) Relation between Ca^{2+}-activated isometric force and muscle fiber stiffness in control muscle fibers (crosses), and that in p-PDM-treated muscle fibers (filled circles). Both force and stiffness are expressed relative to control values as T/T_0 and S/S_0, respectively. Muscle fiber stiffness was measured by applying sinusoidal vibrations (0.1% of fiber length, 1 kHz) and measuring resulting force changes. (Right) Relation between Ca^{2+}-activated isometric force T/T_0 and the maximum unloaded shortening velocity, measured by the slack test, in which muscle fiber was quickly released to be slack, and measuring the time between the quick release and the subsequent force redevelopment (Edman, 1979), and also expressed relative to control value V/V_0. From Chaen, Shimada, and Sugi (1986).

It is of interest that myosin molecules with one native head still hydrolyze ATP at a normal rate, and contribute both muscle fiber stiffness and internal resistance against shortening, being consistent with the idea that they still repeat weak to strong actin binding transition without force generation.

7.4.8 Fiber Stiffness and Internal Resistance

In p-PDM untreated (control) muscle fibers, muscle fiber stiffness, as measured by applying small sinusoidal vibrations (0.1% of fiber length, 1 kHz) to contracting muscle fibers and measuring resulting force changes, increased linearly with increasing magnitude of steady Ca^{2+}-activated isometric force, as shown by the linear stiffness versus force relation (crosses in Fig. 69, left), indicating that the stiffness serves as the measure of the number of myosin heads participating in the force-generating attachment–detachment cycle with actin filaments.

In p-PDM-treated fibers, on the other hand, the curved force versus stiffness relation (filled circles in Fig. 69 left) indicates that muscle fiber stiffness is proportional to the total fraction of native myosin heads ($1 - q$). It follows from this that all native myosin heads hydrolyze ATP with a similar rate, irrespective of whether another myosin head in the same myosin molecule is inactivated by p-PDM or not; in other words, all native myosin heads perform attachment–detachment cycle with actin filaments to keep their normal rate of actin-activated MgATPase activity, irrespective of whether they generate force or not.

We measured the maximum unloaded shortening velocity V_{max} of Ca^{2+}-activated muscle fibers by the slack test (Edman, 1979). The maximum unloaded shortening of p-PDM-treated muscle fibers decreased in parallel with the decreasing isometric force (Fig. 69 right), indicating that myosin molecules, with one head native and the other head inactivated by p-PDM, produce internal resistance against fiber shortening. Namely, if one of the two heads is inactivated by p-PDM, the remaining native head performs attachment–detachment with the actin filament to hydrolyze ATP without generating isometric force.

In contrast, myosin head CAD in myosin molecules, with antibody 3 or anti-S2 antibody attached, maintains its normal actin-activated ATPase activity, despite their supposed weak binding to actin, which is not being sensed by small vibrations as stiffness and produces no appreciable internal resistance against shortening (Figs. 57 and 60). Here, I again emphasize the influence of antibody 3 and anti-S-2 antibody on the strength of binding of myosin head CAD to actin, despite their binding sites distant from myosin head CAD.

Our experiments described above have shown that in muscle, myosin molecules with two native heads can generate force in contracting muscle, while myosin molecules with one native and one inactivated head can no longer generate force. This means that the dogma that each myosin head acts as an independent force generator in the contracting muscle is not exactly correct, though single-headed myosin molecules of S-1 are shown to generate force in in vitro experiments.

The reason why one inactivated head inhibits contractile activity of another native head within the same molecule is not clear and is to be investigated in future together with the possible cooperativity of myosin two heads in producing force and motion in the muscle.

7.4.9 Effect of p-PDM Modification on Myofilament Lattice Structure of Skinned Muscle Fibers

To obtain information about the effect of p-PDM modification of myosin heads on the myofilament lattice structure, skinned muscle fibers were first treated with 200 mM p-PDM to completely eliminate Ca^{2+}-activated force development, and were then quickly frozen to prepare longitudinal and cross sections (Suzuki et al., 1994). Electron micrographs of the longitudinal section of control and p-PDM-inactivated muscle fibers are shown in Fig. 70A,B, respectively. Ladder-like structures between myofilaments, which can be regarded as myosin heads, are more prominent in p-PDM-inactivated fibers than control fibers.

Figure 70C,D shows the Fourier transform diffractograms of the myofilament overlap region (indicated by square-shaped areas in Fig. 70A,B). The diffractograms obtained from p-PDM-inactivated fibers exhibited distinct meridional reflections with axial spacing of 14.3 nm (Fig. 70D), which arises from the axial spacing between adjacent myosin heads on the myosin filament, while in the diffractograms from control fibers the 14.3 nm meridional reflections were not observed (Fig. 70C). These results indicate that in p-PDM-inactivated fibers, the regularity of myosin head orientation on the myosin filament increases compared to that in control fibers. It is of interest that, in quickly frozen replicas of muscle fibers, myosin head angular distribution always shows a peak around 90° in all the states examined (Figs. 33 and 34). This strongly suggests that p-PDM-inactivated myosin heads extend from myosin filaments at an angle of ~90°. This again supports the view that myosin heads detached from actin filaments take their neutral (equilibrium) position, at which they extend from myosin filaments at an angle of 90°. This point will be discussed again in connection with our results obtained from the EC experiments on ATP-induced myosin head power stroke.

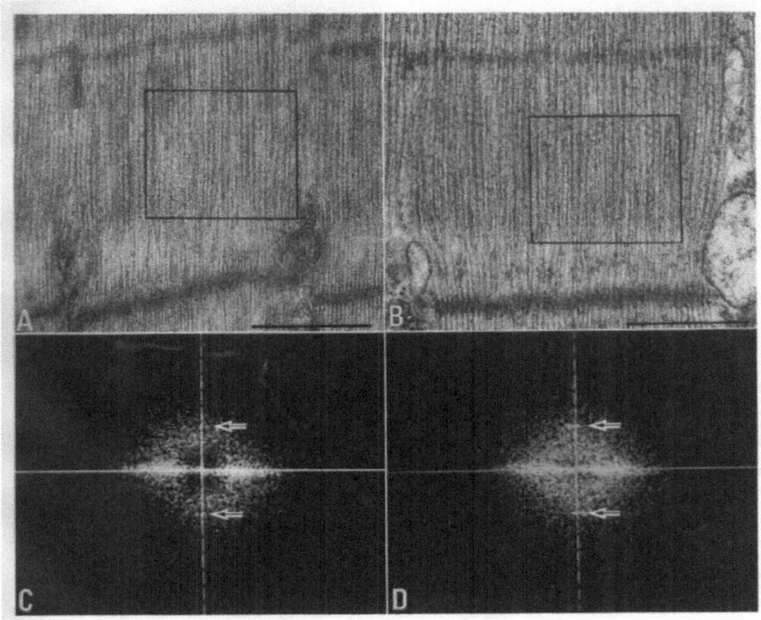

Figure 70 Longitudinal sections of quickly frozen control (A) and p-PDM-inactivated (B) skinned muscle fibers, and their Fourier transform diffractograms (C and D) obtained from respective square-shaped areas. Note that ladder-like structures are more prominent in B than in A, and that clear 14.3 nm meridional reflections (arrows) can be seen in D but not in C. Bar, 0.5 μM. Suzuki et al. (1994).

We also examined cross sections of control and p-PDM-inactivated fibers on the regularity of hexagonal myofilament lattice structure and found no appreciable effect of p-PDM on it.

Chapter 8

Electron Microscopic Recording of Myosin Head Power Stroke: Evidence That Myosin Heads Take the Neutral Position When Detached from Actin Filaments

In this chapter, I will describe our success in recording myosin head power stroke associated with ATP hydrolysis, using a mixture of actin and myosin filaments. In relaxed muscle, myosin heads are believed to be in the form of complex M-ADP-Pi (Lymn and Taylor, 1971), while in the actin and myosin filament mixture, myosin heads form rigor linkages A-M with actin filaments.

It will be shown that the ATP-induced myosin head power stroke in the filament mixture is essentially the same as that taking place in contracting muscle, despite the above difference in the initial condition of myosin heads, if it is assumed that myosin heads detached from actin filaments always take a configuration, i.e., neutral position, perpendicular to myosin filament axis. This idea can explain all the results obtained from our EC experiments with respect to ATP-induced myosin head position changes in both the absence and the presence of actin filaments.

Mysteries in Muscle Contraction: Evidence against Current Dogmas
Haruo Sugi

ISBN 978-981-4774-48-2 (Hardcover), 978-1-315-10191-0 (eBook)
www.panstanford.com

8.1 Preparation of Actin and Myosin Filament Mixture

To record the ATP-induced myosin head power stroke in the presence of actin filaments, we prepared a mixture of actin and myosin filaments by mixing bipolar synthetic myosin filaments (Fig. 31) with actin filaments, in such a way that actin filaments were in large excess of myosin filaments. Conventional electron micrographs of the actin and myosin filament mixture are shown in Fig. 71. Although the filament mixture tended to aggregate, we occasionally found spindle-shaped myosin filaments surrounded by a number of actin filaments, running not only in random directions but also parallel to the myosin filament long axis.

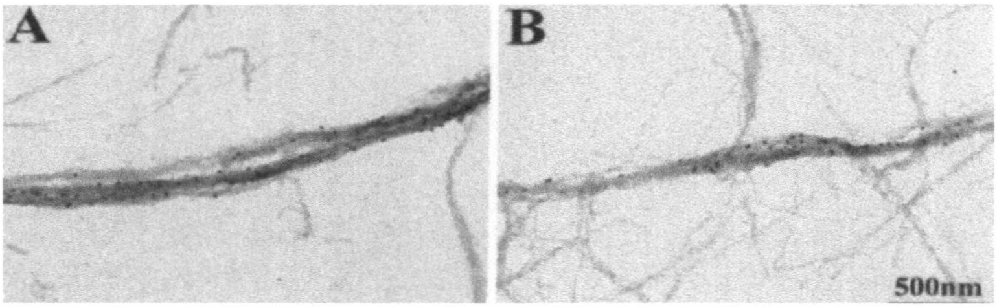

Figure 71 Conventional electron micrographs of actin and myosin filament mixture. Myosin heads are position marked with antibody 1 in A, and with antibody 2 in B. Note myosin filaments are surrounded by actin filaments. From Sugi et al. (2015).

Before ATP application, myosin heads are expected to form tight rigor linkages with actin filaments. This expectation was supported by the stability in position of individual myosin heads, which was much more pronounced compared to that observed in myosin filaments in the absence of actin filaments (Sugi et al., 2008). The change in the center of mass position of individual myosin heads, as measured by taking two IP records of the same filament at intervals of 5–10 min was 0.6 ± 0.06 nm (mean ± SD, $n = 220$), indicating myosin heads actually form tight rigor linkages with actin filaments. This provided favorable conditions to measure the ATP-induced myosin head power stroke, which was expected to be very small if a limited proportion of myosin heads were activated with a limited amount of ATP is taken into consideration.

In the previous experiments performed in the absence of actin filaments (Sugi et al., 1997, 2008; Minoda et al., 2011), it has been shown that myosin head thermal fluctuation occurs around the definite equilibrium position, in accordance with the assumption in the Huxley contraction model (A. F. Huxley, 1957). It may follow from this that when myosin filaments are mixed with actin filaments, most myosin heads may form rigor linkages with actin at or near their equilibrium position, where myosin heads produce no force. Since the rigor actin–myosin linkage has a finite lifetime depending on the load on it (Nishizaka et al., 2000), rigor linkages, which are occasionally formed distant from the equilibrium position, would be broken after some period of time to be replaced by those formed at or near the equilibrium position during the time from the filament mixing to the beginning of experiments (30~40 min).

8.2 Amplitude of ATP-Induced Myosin Head Power Stroke

As shown in Fig. 72, individual myosin heads in the filament mixture were observed to move in response to ATP application.

Reflecting the very limited proportion of myosin heads activated by applied ATP, the amplitude of the myosin head was much smaller than that of the myosin head recovery stroke in the absence of actin filaments (Sugi et al., 2008; Minoda et al., 2011). As most myosin heads in the filament mixture form rigor linkages with actin filaments, a small proportion of myosin heads activated by ATP cannot produce gross myofilament sliding and can only stretch adjacent sarcomere structures for a limited distance. This condition is analogous to the isometric condition in muscle, in which a muscle is made to contract with both ends fixed in position. In an isometrically contracting muscle, however, "strong" sarcomeres producing more force than "weak" sarcomeres shorten internally by stretching "weak" sarcomeres, thus producing longitudinal non-uniformity in the sarcomere length.

Since both vertical and horizontal axes of a pixel (2.5 × 2.5 nm under electron microscopic magnification of 10,000×) were divided into two significant figures, it was possible to measure the amplitude of power stroke with a spatial resolution up to 0.25 nm.

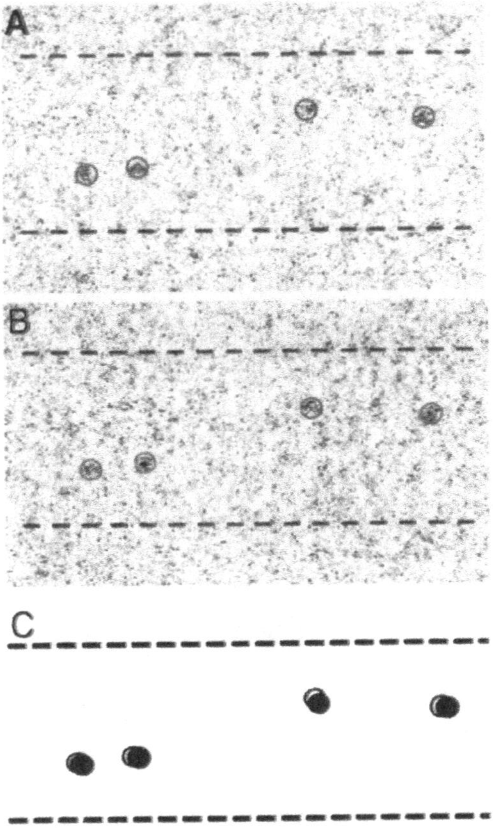

Figure 72 A pair of IP records showing part of the same myosin filament with gold particles attached to myosin heads via antibody 1. Records A and B were taken before and after ATP application, respectively. Circles (diameter, 20 nm; red in A and blue in B) are drawn around the center of mass position of the same particles attached to individual myosin heads before and after ATP application. In C, open and filled circles are drawn around the center of mass positions of the same particles, before and after ATP application, respectively (Sugi et al., 2015).

In the synthetic myosin filaments used in our EC experiments, more or less similar non-uniformity may be expected to take place; namely, a region within the myosin filament, where the number of ATP-activated myosin heads is larger than that within adjacent regions, may stretch the "weak" adjacent region to a certain extent.

This may cause distortion of the filament structure, reversibly or irreversibly to more or less make the amplitude of the power stroke, recorded on the IP ambiguous.

Figure 73 shows histograms of amplitude distribution of myosin head power stroke. The average power stroke amplitude was larger in the distal region of myosin head CAD, position-marked with antibody 1, than in the proximal region of myosin head CAD, position-marked with antibody 2. The average amplitude of myosin head power stroke was 3.3 ± 0.2 nm (mean ± SD, n = 732) at the distal region, and 2.5 ± 0.1 nm (mean ± SD, n = 613) at the proximal region of myosin head CAD. The small values of the myosin head power stroke with a small range of variation indicates that the distortion of the myosin filament structure is very small; otherwise, the power stroke amplitude may show much wider variation. This was further confirmed by experiments that will be described in the next section.

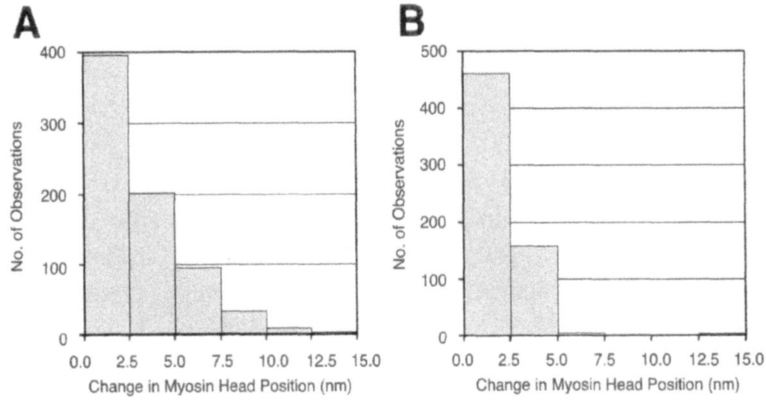

Figure 73 Histograms showing amplitude distribution of ATP-induced myosin head power stroke at standard KCl concentration (120 mM) in experimental solution. (A) Amplitude distribution of myosin heads position-marked with antibody 1 attaching the distal region of myosin head CAD. (B) Amplitude distribution of myosin heads position-marked with antibody 2 attaching the proximal region of myosin head CAD. From Sugi et al. (2015).

8.3 Reversibility of ATP-Induced Myosin Head Power Stroke

To ascertain whether the ATP-induced myosin head power stroke is reversible, we took three IP records of the same filament; (1) before ATP application, (2) 40~60 s after the onset of current pulse to ATP-containing microelectrode, and (3) 5~6 min after ATP application, i.e., after complete exhaustion of applied ATP in experimental solution containing hexokinase and D-glucose serving as ATP scavenger (Oiwa et al., 1991, 1993).

Figure 74 presents examples of IP records showing sequential changes in position of pixels (2.5 × 2.5 nm), in which the center of mass positions of corresponding gold particles were located. Pixel positions in the first, second, and third records are colored red, blue, and yellow, respectively. Similar results were obtained irrespective of whether myosin heads were position-marked with antibody 1 or antibody 2.

When the amplitude of power stroke was <5 nm, myosin heads returned almost exactly to their initial position before ATP application, as indicated by the complete overlap of the pixel position in the first and the third records, so that yellow pixels were partly or completely covered by red pixels (Figs. 74A–C). With larger amplitudes of ATP-induced myosin head movements, myosin heads tended to return to the position close to their initial position before ATP application (Figs. 74D,E,F). These results indicate that the expected distortion of the myosin filament structure caused by non-uniform distribution of ATP-activated myosin heads within the filament is rather small, so that the reversible myosin head movements with small amplitudes can be clearly recorded as shown in Fig. 74.

As already discussed, before ATP application, individual myosin heads are believed to form rigor linkages with actin filaments at their equilibrium (or neutral) position, at which myosin heads generate no force (A. F. Huxley, 1957). The nearly exact return of individual myosin heads to their initial position after the exhaustion of applied ATP may indicate that after complete exhaustion of applied ATP, myosin heads again form rigor actin–myosin linkages at their neutral position, if the return of myosin heads after detachment from actin filaments is assumed to occur rapidly. This point is important in constructing an idea about the sequential events in myosin head–actin attachment–detachment cycle producing muscle contraction to be discussed later.

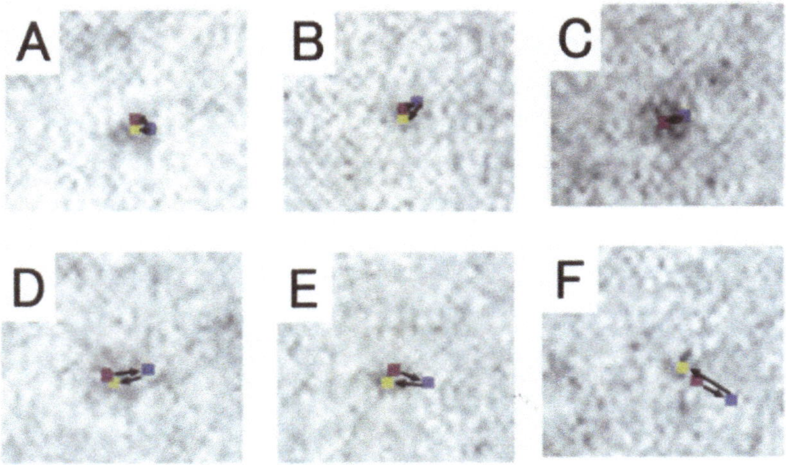

Figure 74 Sequential changes in position of different pixels (2.5 × 2.5 nm) where the center of mass positions of corresponding gold particles are located. In each frame, pixel positions were recorded three times, i.e., before ATP application (red), after ATP application (blue), and after complete exhaustion of applied ATP (yellow). Direction of each particle movement, i.e., direction of myosin head movement, is indicated by arrows. Frames (A, C, D) were obtained from myosin heads position-marked with antibody 1, while frames (B, E, F) were obtained from myosin heads position-marked with antibody 2. From Sugi et al. (2015).

8.4 Enhanced Performance of Myosin Heads at Low Ionic Strength: Agreement of Results between EC and Physiological Experiments

8.4.1 Physiological Experiments

It has been known that isometric force generation in Ca^{2+}-activated skinned muscle fibers is markedly enhanced at low ionic strength (Gordon et al., 1973; Thames et al., 1974). By simultaneously recording isometric force development and MgATPase activity of skinned muscle fibers, we have shown that when KCl concentration in the experimental solution is reduced from 125 to 0 mM (corresponding to decrease of ionic strength, μ, from 170 mM to 50 mM), the magnitude of Ca^{2+}-activated isometric force increases about twofold, while the MgATPase activity does not change appreciably (Fig. 75) (Sugi et al., 2013).

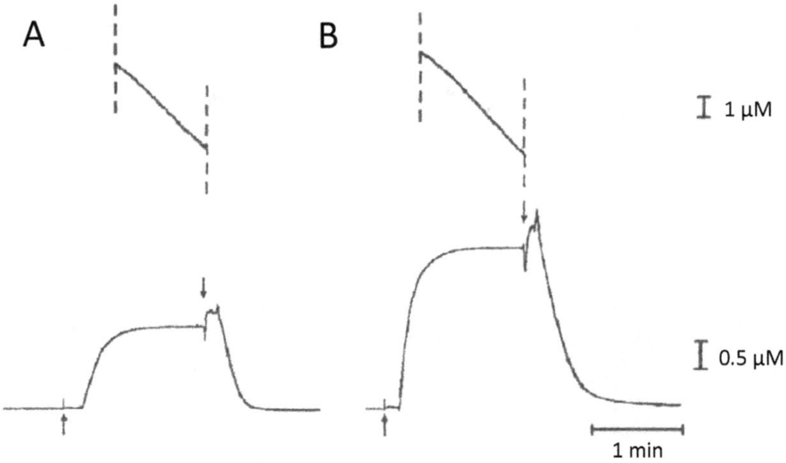

Figure 75 MgATPase activity (upper traces) and Ca^{2+}-activated isometric force (lower traces) in skinned muscle fibers at 125 mM KCl (A) and at 0 mM KCL (B). Note that the slope of MgATPase records does not change appreciably, while isometric force increases ~twofold at low ionic strength. From Sugi et al. (2013).

Force–velocity relations of Ca^{2+}-activated muscle fibers at various ionic strengths are shown in Fig. 60. It can be seen that despite variation of the maximum isometric force depending on ionic strength of experimental solution, the maximum unloaded velocity remains unchanged (Fig. 76A). If force–velocity curves were normalized with respect to the maximum isometric forces, the curves were found to be identical in shape (Fig. 76B).

The results shown in Figs. 75 and 76 can be most readily accounted for by assuming that the increase in Ca^{2+}-activated isometric force results from the increased ability of individual myosin heads to generate force without changing their kinetic properties, but not from increase in the number of myosin heads participating in force generation. This implies that at low ionic strength, the amplitude of individual myosin heads in the isometric condition may increase as they can stretch adjacent structures with their increased forces.

As described in the next section, the result of the EC experiments, that the amplitude of ATP-induced power stroke in individual myosin heads also increases nearly twofold at low ionic strength, is completely consistent with the results obtained from the physiological experiments.

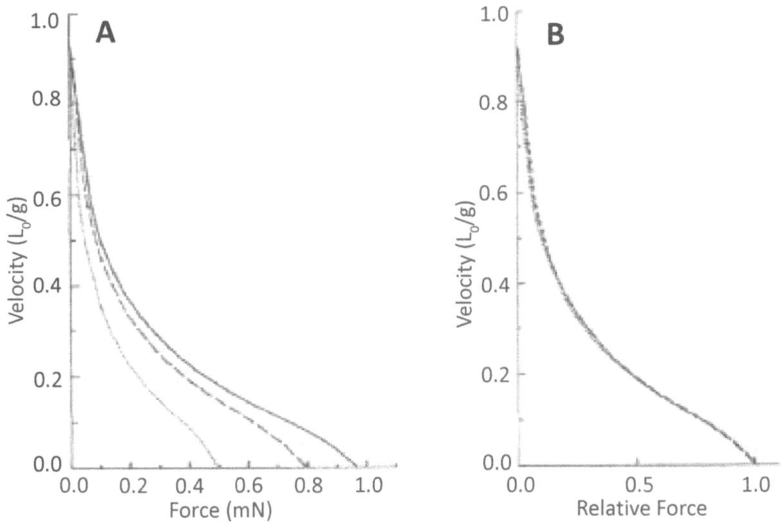

Figure 76 (A) Force–velocity curves obtained at 0 mM KCl (solid line), at 50 mM KCl (broken line) and at 125 mM KCl (dotted line). Note that the maximum unloaded shortening velocity remains unchanged despite marked decrease of the maximum isometric forces. (B) Force velocity curves normalized with respect to the maximum isometric force. Note that the curves are identical in shape. From Sugi et al. (2013).

8.4.2 EC Experiments

Figure 77 is an example of a pair of IP records of the same filaments taken before (A) and after (B) application of ATP, together with diagram showing the ATP-induced myosin head movement as open and filled circles, drawn around the center of mass positions of the same particles before and after ATP application, respectively (C).

Some gold particles move in the direction nearly normal to the filament axis probably due to irregular arrangement of actin filaments around myosin filaments. As shown in Fig. 78, histograms of amplitude distribution of the myosin head power stroke indicated that, at both distal and proximal regions of myosin head CAD, the amplitude of ATP-induced myosin head movement was nearly the same and obviously larger at low ionic strength than at standard ionic strength (Figs. 73 and 78). The features of myosin head power stroke was the same as those of myosin head recovery stroke (Fig. 48).

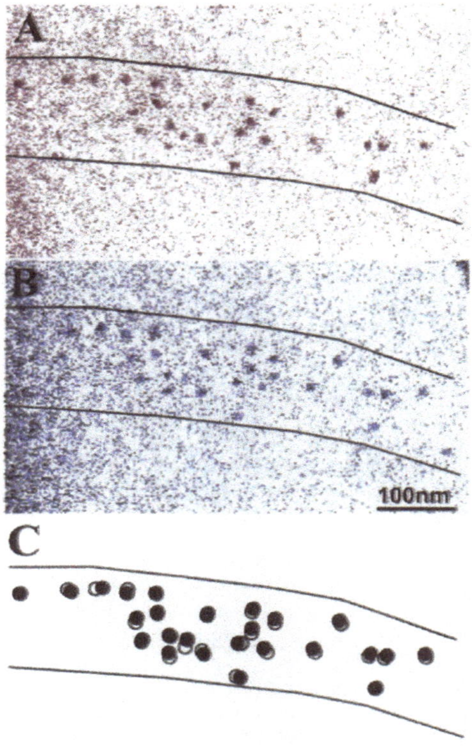

Figure 77 A pair of IP records of the same myosin filament in filament mixture taken before ATP application (A, colored red)) and after ATP application (B, colored blue). (C) Diagram showing ATP-induced changes in position of gold particles attached to individual myosin heads with antibody 2. Open and filled circles (diameter, 20 nm) are drawn around center of mass positions of the same particle before and after ATP application, respectively. From Sugi et al. (2015).

The amplitude of power stroke at the distal region of myosin head CAD was 4.4 ± 0.1 nm (mean ± SD, n = 361), and 4.3 ± 0.2 nm (mean ± SD, n = 305) at the proximal region of myosin head CAD. Thus, the amplitude of power stroke did not differ significantly between the distal and the proximal regions of CAD, being consistent with the prediction of swinging lever arm hypothesis that, during the power stroke, myosin head CAD moves along the actin filament without changing its attachment angle (Fig. 48), though this does not necessarily mean that this hypothesis is correct. The swinging lever arm hypothesis is constructed from extracted protein samples and therefore bears no direct relation to our EC experiments at all.

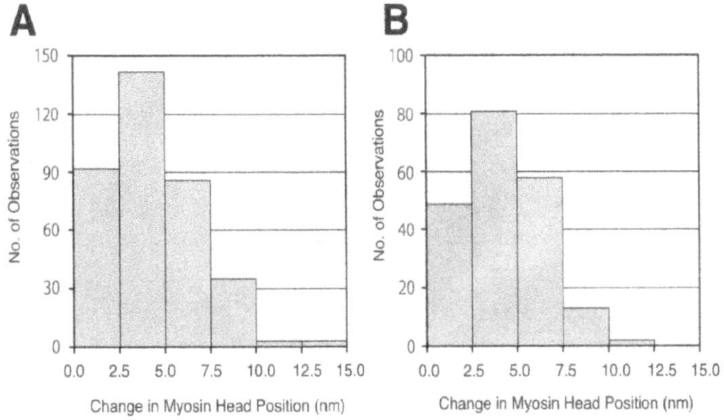

Figure 78 Histograms showing amplitude distribution of ATP-induced myosin head power stroke at low ionic strength (external KCl concentration, 20 mM). Individual myosin heads were position-marked with antibody 1 attaching to the distal region of myosin head CAD in A, and with antibody 2 attaching to the proximal region of myosin head CAD in B. From Sugi et al. (2015).

Meanwhile, the increase in the amplitude of myosin head power stroke at low ionic strength is completely consistent with results of physiological experiments (Figs. 75 and 76) that, at low ionic strength, the force generated by individual myosin heads increases about twofold; in the condition in which gross myofilament sliding does not take place in the filament mixture, each ATP-activated myosin head only exerts power stroke by stretching adjacent structures, and the extent of stretch increases with increasing force in each myosin head.

8.5 Two Different Modes of Myosin Head Power Stroke As Revealed by EC Experiments

Figure 79 illustrates two different modes of myosin head power stroke, as revealed by our EC experiments. The most prominent advantage of EC experiments over any other experiments is the capability to record the amplitude of myosin head movement at different regions within a myosin head. By using three different antibodies, we have already shown that during myosin head recovery stroke, the amplitude of movement at the distal region of myosin head CAD is the same as that at the proximal region of myosin head CAD (Fig. 48) (Minoda et al., 2011).

In the actin and myosin filament mixture, where only a small proportion of myosin heads can be activated by ATP, we have shown that the amplitude of power stroke is larger at the distal region of myosin head CAD than at its proximal region at standard ionic strength (Fig. 73), indicating that the attachment angle of myosin head CAD is not always perpendicular to the actin filament axis.

At low ionic strength, on the other hand, the amplitude of myosin head power stroke is the same between the distal and the proximal regions of myosin head CAD (Fig. 78), so that the attachment angle of myosin head CAD is perpendicular to actin filament axis during the power stroke. This result, together with the results with respect to myosin head recovery stroke (Fig. 48), suggests that the amplitude of both power and recovery strokes are the same when myosin heads can move freely in the absence of actin filaments, or can generate strong force to stretch adjacent elastic structures for a distance ≥ 4 nm (Fig. 78).

It should again be noted that, though EC experiments are consistent with the prediction of the swinging lever arm hypothesis that myosin head CAD is rigid and moves along the actin filament without changing its attachment angle (Figs. 48 and 78), it does not mean that power stroke is produced only by the active rotation of LD around COD. As already mentioned several times, the swinging lever arm hypothesis is based on the static structure of the truncated myosin crystal, and ignores the essential role of myosin head LD and myosin S-2. Consequently, the agreement between the results of EC experiments and the prediction of swinging lever arm hypothesis is just accidental.

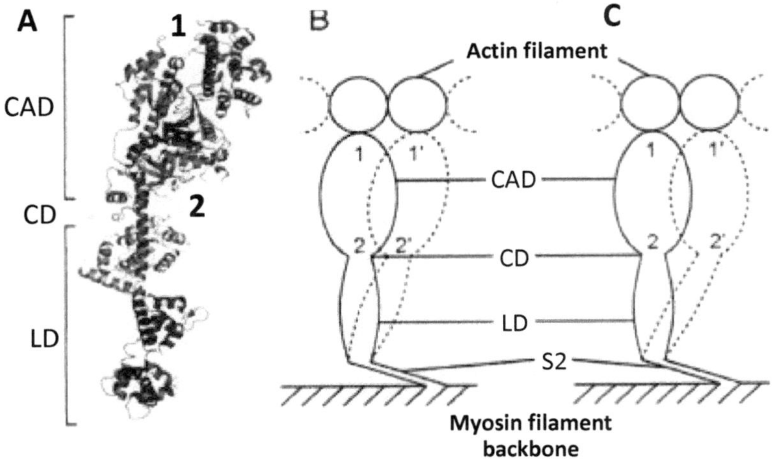

Figure 79 Change in the mode of myosin head power stroke in the isometric condition. (A) Diagram of myosin head structure consisting of catalytic domain (CAD), converter domain (COD), and lever arm domain (LD). Approximate attachment regions of antibody 1 and antibody 2 are indicated by numbers 1 and 2, respectively. (B) the mode of myosin head power stroke at standard ionic strength. The amplitude of movement is larger at the distal region than at the proximal region of myosin head CAD. (C) The mode of myosin head power stroke at low ionic strength. The amplitude of movement is the same at both the distal and the proximal regions of myosin head CAD. From Sugi et al. (2015).

8.6 Sequence of Changes in Myosin Head Configuration during ATP-Induced Myosin Head Power and Recovery Strokes

In this section, I will explain that the sequence of changes in myosin head configuration taking place during the ATP-induced myosin head power stroke in the filament mixture is essentially the same as those in the myosin head power stroke in contracting muscle, based on the assumption that the individual myosin heads take their equilibrium (neutral) position if they are detached from actin filaments (Fig. 6; A. F. Huxley, 1957). In the filament mixture, myosin heads may form rigor linkages with actin filaments, as supported by the marked stability of myosin head position before ATP application (Sugi et al., 2015) compared to that of myosin heads in the absence of actin filaments (Sugi et al., 2008). Myosin heads (M) in the filament mixture are therefore in the state of A-M (corresponding to B, in Fig. 80).

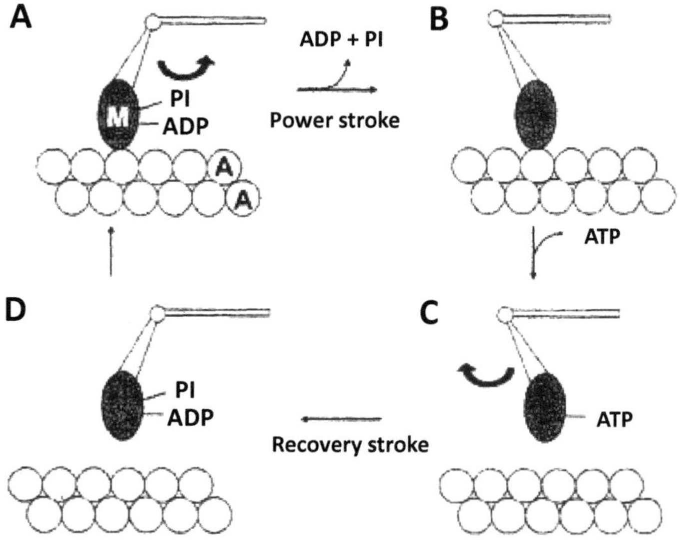

Figure 80 Comprehensive putative diagram of attachment–detachment cycle between myosin head (M) extending from myosin filament and actin (A) in actin filament. M in the form of M-ADP-Pi attaches to A to exert power stroke, associated with release of Pi and ADP (From A to B). After completion of power stroke, M remains attached to A, taking its post-power stroke configuration (B). Upon binding with next ATP, M detaches from A (B to C) to exert recovery stroke, associated with ATP hydrolysis to form complex M-ADP-Pi (from C to D). Then M-ADP-Pi again attaches to A to again exert power stroke (from D to A). In this diagram, M is assumed to be rigid, and moves parallel to actin filament without changing its attachment angle to A. From Sugi et al. (2008).

8.6.1 Myosin Heads Take Neutral Position Where They Exert No Force When Detached from Actin Filaments

In relaxed muscle, on the other hand, M is in the state of M-ADP-Pi (corresponding to D, in Fig. 80). The above difference in the initial state of M between the filament mixture and the relaxed muscle may give an impression that the ATP-induced myosin head power stroke in the filament mixture is qualitatively different from myosin head power stroke in muscle. As explained below, the condition of myosin heads before ATP application is essentially similar to that in relaxed muscle: if myosin heads are assumed to take a neutral (or equilibrium) position where they exert no force in the absence of actin filaments or when detached from actin filaments (A. F. Huxley, 1957; Fig. 7).

Hereafter, I will use the word "neutral" instead of "equilibrium" for the following reasons: (1) In Fig. 6A, myosin head is connected to the myosin head backbone via two springs, giving readers an impression that myosin heads fluctuate only on a single plane parallel to myosin filament axis. (2) Myosin heads in myofilament lattice may fluctuate not only on a specific plane but also in all directions around a definite central position. (3) Consequently, I prefer "neutral" to "equilibrium", since the former word expresses more accurately myosin head movements in all directions.

In Fig. 80, diagram B represents A-M at the end of myosin head power stroke, i.e., in the force-generating configuration until next ATP comes to bind with M. This scheme applies when the cyclic A-M interaction goes on rapidly at high ATP concentrations in contracting muscle. The situation in the filament mixture before ATP application is different from that in muscle. In the filament mixture, myosin heads slowly repeat formation and breaking if rigor A-M linkges to their finite lifetime (30 s to several minutes; Nishizaka et al., 2000).

The process of formation and breaking of A-M linkages is shown in Fig. 81, where myosin heads are shown as oval-shaped heads (M) connected to the myosin filament by vertical or oblique lines. Myosin heads, taking their neutral position, where they generate no force, are colored blue and connected to the myosin filament with vertical lines, while myosin heads taking positions distant from their neutral position and therefore exerting rigor force in random directions are colored red and connected to the myosin filament with oblique lines.

In the filament mixture, rigor A-M linkages are formed by myosin heads either at neutral position (blue) or at the non-neutral position (red) in the absence of ATP (Fig. 81A). When

A-M linkages are broken occasionally, myosin heads (M) quickly return to their neutral position before again attaching to actin filaments (Fig. 81B to C). After taking the neutral position, each myosin head again forms rigor A-M linkages by binding with actin monomers, just facing it, i.e., an actin monomer nearest to each myosin head (Fig. 81B to D). Thus, rigor A-M linkages generating rigor force are expected to be gradually replaced by those generating no force.

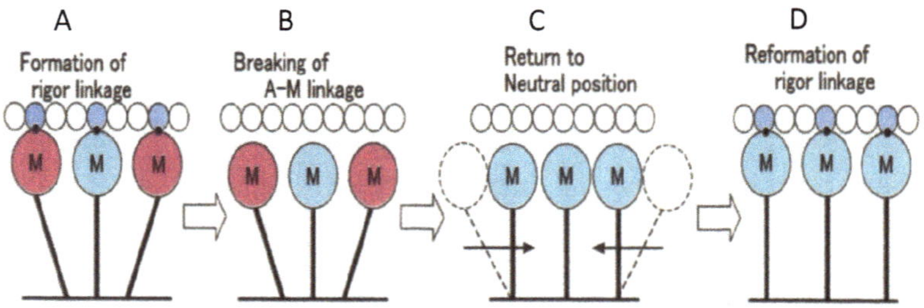

Figure 81 Diagrams illustrating breaking and reformation of rigor A-M linkages. Myosin heads (M) taking neutral (equilibrium) position are colored blue with vertical lines, while myosin heads not taking neutral position are colored red with oblique lines. For further explanation, see text.

Since it takes ~30 min to start EC experiments after the preparation of the filament mixture, almost all myosin heads may form rigor linkages with actin filaments at their neutral position. On application of ATP, myosin heads at their neutral position detach from actin filaments to form M-ADP-Pi. This condition is the same as that when relaxed muscle is activated to contract.

8.6.2 Why Only Two Definite Myosin Head Positions Are Recorded in EC Experiments

In this section, I will explain the reason why only two distinct myosin head positions are recorded with respect to both ATP-induced recovery and power strokes (Figs. 47 and 74) in EC experiments, based on the assumption that myosin heads take their neutral, non-force-generating configuration when they are detached from actin filaments (A. F. Huxley, 1957).

8.6.2.1 ATP-induced myosin head recovery stroke in the absence of actin filaments

As illustrated diagrammatically in Fig. 82, a myosin head (M, colored blue, connected to myosin filament with vertical line) maintains its neutral position in the absence of ATP, so that it exhibits a stable time-averaged position despite thermal fluctuations (A). Thus, the stable position of M before ATP application can be recorded on the IP. When externally applied ATP comes to bind with myosin head (A to B), it exerts recovery stroke, associated with reaction, M + ATP → M-ATP → M-ADP-Pi in the absence of actin filaments (C). As M-ADP-Pi (M, colored red) has a long average lifetime > 10 s, the myosin head (M, colored red, connected to myosin filament with oblique line) taking the post-recovery stroke configuration can also be recorded on the IP.

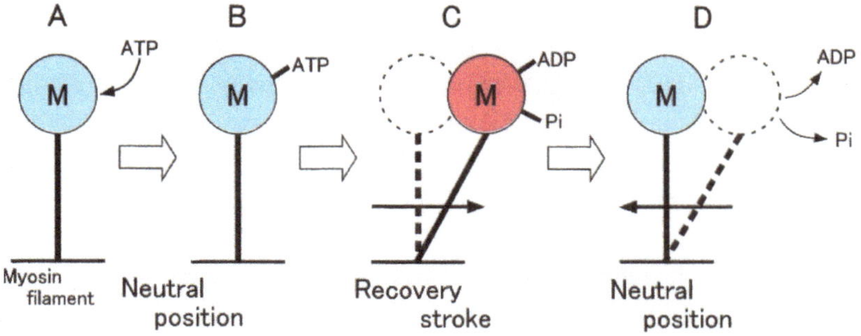

Figure 82 Sequence of myosin head movement during ATP-induced recovery stroke in the absence of actin filaments. In the absence of actin filament, myosin head at its neutral position binds with externally applied ATP (A to B) and exerts recovery stroke, associated with ATP hydrolysis (B to C). After releasing Pi and ADP, myosin head returns to its neutral configuration (C to D).

Although myosin head in the state of M-ADP-Pi is charged up with chemical energy of ATP hydrolysis, the energy is dissipated after the release of Pi and ADP. After losing the energy, myosin head returns to its initial neutral configuration, which can also be recorded on the IP after exhaustion of externally applied ATP. As a result, the position of the myosin head is completely or nearly the same between the IP records taken before ATP application and taken after the exhaustion of ATP.

As demonstrated in Fig. 46, myosin heads exhibit ATP-induced movement away from the bare region, i.e., recovery stroke in the absence of ATP. This finding indicates that myosin heads have a basic property that, if they do not bind with actin filaments, move away from the bare region to perform recovery stroke, associated with reaction, M-ATP → M-ADP-Pi. Conversely, when myosin heads attach to actin filaments, they move toward the bare region to perform power stroke, associated with reaction, A-M-ADP-Pi → A-M + ADP + Pi. This implies that myosin heads can perform recovery stroke without being guided by actin filaments; in other words, to perform recovery stroke is a property inherent to myosin heads detached from actin filaments or in the absence of actin filaments.

8.6.2.2 ATP-induced myosin head power stroke in the filament mixture

Diagrams illustrating the sequence of myosin head configuration changes during the ATP-induced power stroke in the presence of actin filaments in the EC experiments are presented in Fig. 83. Before ATP application, myosin head forms rigor linkage with actin filament in its neutral position (M, colored blue, connected to myosin filament with vertical line) (A), as already explained in Fig. 67. On binding with ATP, it detaches from actin filament (B) and attaches again to the actin filament in the charged up state, M-ADP-Pi (M, colored red) (C). Then myosin head performs power stroke by utilizing chemical energy for mechanical work producing myofilament sliding (M, colored red, connected to myosin filament by oblique line, i.e. force-producing configuration) (D). At the end of power stroke, myosin head remains attached to actin filament by forming rigor linkage until next ATP comes to bind it (M, colored blue) (E).

In the EC experiments, it takes time for the myosin head to find the next ATP, because of low ATP concentrations around the myosin filament, so that M is in the state of A-M for a considerable time (1 to several s) to be recorded by the IP system. This constitutes the reason why ATP-induced myosin head movement can be recorded by the low time resolution (0.1 s) of IP recording. On binding with the next ATP, myosin head detaches from actin filament to quickly restore its neutral configuration (F) and again attaches to actin filament in the charged up state of M-ADP-Pi (G) (M, colored red, connected to myosin filament by a vertical line). Thus, in the EC experiments, we record two distinct myosin head position, i.e., the position before ATP application (A) and the position after completion of the power stroke (E), as has been shown in Figs. 47 and 74.

As described above, the results of EC experiments with respect to both power and recovery strokes can be accounted for by assuming the property of myosin heads to take their neutral position if they are detached from actin filaments. More or less similar sequence of events may take place in optical trap experiments, since the records with high trap stiffness consist of transitions between non-force-generating and force-generating states (Fig. 25).

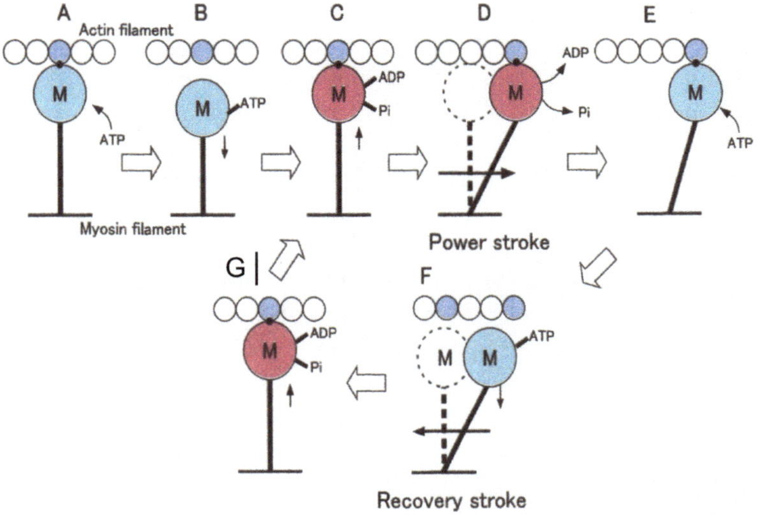

Figure 83 Diagrams illustrating the sequence of events during ATP-induced myosin head power and recovery strokes in the presence of actin filaments. A myosin head at its neutral position (M, colored blue) binds with externally applied ATP (A to B). The myosin head charged up with chemical energy of ATP hydrolysis (M, colored red) attaches to actin filament in the form of M-ADP-Pi (C) to exert power stroke (M, colored red) (D). After completion of power stroke,, associated with release of Pi and ADP, the myosin head (M, colored blue) stays attached to actin filament (E). On binding with next ATP, M (colored blue) detaches from actin filament to return to its neutral position (F), and then again attaches to actin filament in the charged up M-ADP-Pi state (G).

In the optical trap experiments, myosin head fixed on the bead forms rigor linkage with actin filament, which is brought into contact with it. At micromolar ATP concentrations, the myosin head repeats the attachment–detachment cycle with actin filament, and in each cycle the myosin head would rapidly return to its neutral configuration when detached from actin filament, a condition similar to that illustrated in Fig. 84.

8.7 Possible Electrostatic Interaction between Myosin Head and Actin Filaments

As has been the case in muscle contraction and in vitro actin–myosin sliding (Figs. 49 and 50), myosin heads, position-marked with antibody 1, exhibited power stroke in response to ATP (Figs. 55–57, 60–61). Since antibody 1, attached to the junctional peptide between 50K and 20K fragments of myosin heavy chain, covers two main actin-binding sites on myosin CAD (Fig. 20) to inhibit the formation of "rigor" complex A-M, our success in recording the myosin head power stroke with the EC again indicates that during muscle contraction, myosin heads do not pass through rigor configuration as determined by extracted protein samples (Fig. 19). In this connection, we have already shown that in the rigor state produced by the removal of external ATP, myosin heads are mostly in the state of A-M-ADP with somewhat dynamic properties (Fig. 58; Sugi et al., 2016).

In this connection, I want to mention accumulating evidence that the linkage between myosin head and action filament in contracting muscle is maintained by an electrostatic mechanism. It has long been recognized that the X-ray diffraction pattern from contracting muscle is intermediate between those from relaxed and rigor muscles, though a considerable proportion of myosin heads in contracting muscle are expected to be strongly bound to actin filaments in the form of A-M ADP and A-M (Squire, 1981).

8.7.1 Anisotropic Nature of Contracting Muscle as Revealed by the Use of Ultrasonic Waves

To obtain information about myosin heads tightly bound to actin filaments during the course of contraction and relaxation, we measured muscle stiffness in both longitudinal and transverse directions (Tamura et al., 1982; Hatta et al., 1988). We have developed a technique in which muscle stiffness changes are continuously recorded by measuring propagation velocity of ultrasonic waves by applying negligibly small perturbations to the contractile system with high time resolution. To measure muscle stiffness in the longitudinal direction, a frog muscle (M) was held by a Lucite block with a cylindrical groove and bent around the block near its both ends. Two ceramic transducers were put in contact with the muscle; one transmitted ultrasonic waves to muscle, while the other sensed the propagated waves (Fig. 85A). To measure muscle stiffness in the long axis, muscle (M) was placed with its lower surface in contact with the bottom of the experimental chamber (G), serving as the reflecting plane of ultrasonic waves (MHz region, Fig. 85B). In both cases, the muscle was fixed in position at one end, while the other end was connected to a force transducer (F).

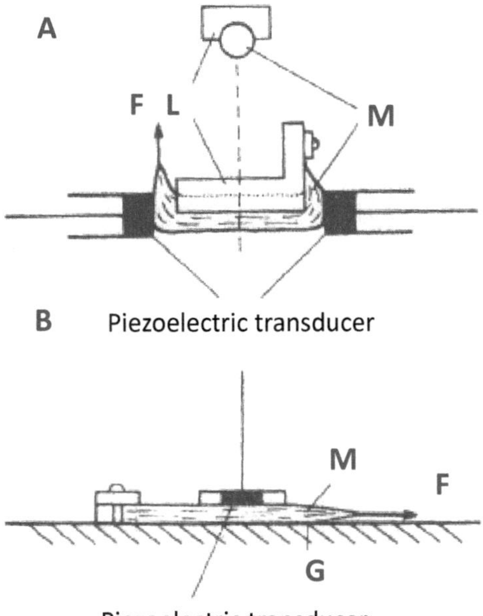

Figure 85 Measurement of muscle stiffness in the longitudinal direction (A) and in the transverse direction (B). In (A), two ceramic (piezoelectric) transducers were in contact with two bent muscle surface, while in (B), muscle was placed horizontally on the surface of experimental chamber, serving as reflecting plane for ultrasonic waves transmitted from upper ceramic transducer. From Tamura et al. (1982).

The results are shown in Figs. 86 and 87. Muscle longitudinal stiffness increased with increasing isometric force. In contrast, muscle transverse stiffness was found to decrease with increasing isometric force. Both the longitudinal and the transverse stiffness changes preceded isometric tension development, indicating that the stiffness changes reflect the behavior of myosin heads preceding the process of isometric force generation.

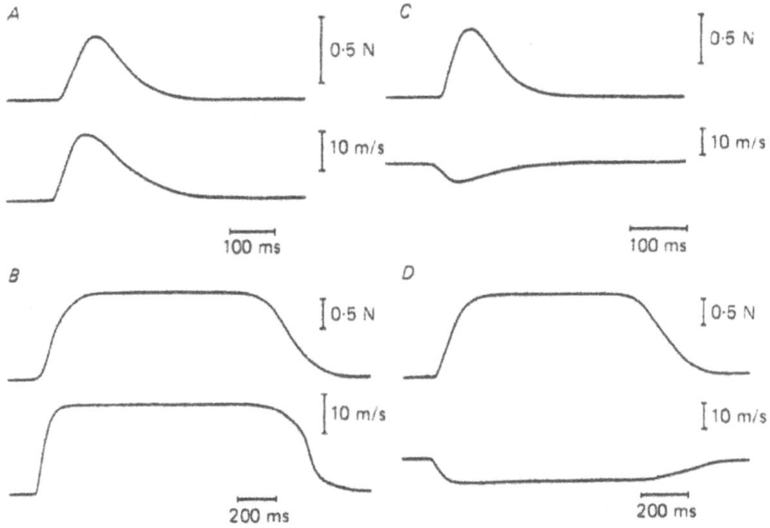

Figure 86 Simultaneous recordings of isometric force (upper traces) and muscle stiffness (lower traces). Changes in the longitudinal muscle stiffness during twitch and tetanus are shown in A and B, respectively. Changes in the transverse muscle stiffness during twitch and during tetanus are shown in C and D, respectively. In all stiffness records, upward and downward changes indicate increase and decrease of stiffness, respectively. From Hatta et al. (1988).

When a muscle was put into rigor state, the stiffness increased in both longitudinal and transverse directions, indicating that the marked anisotropic nature of stiffness contracting muscle reflects dynamic nature of actin–myosin linkages. Actin–myosin linkages in rigor muscle are tight chemical linkages, while actin–myosin linkages, formed during cyclic attachment–detachment cycle in contracting, may not be simple chemical linkages, but some other kind of linkages with anisotropic and dynamic properties.

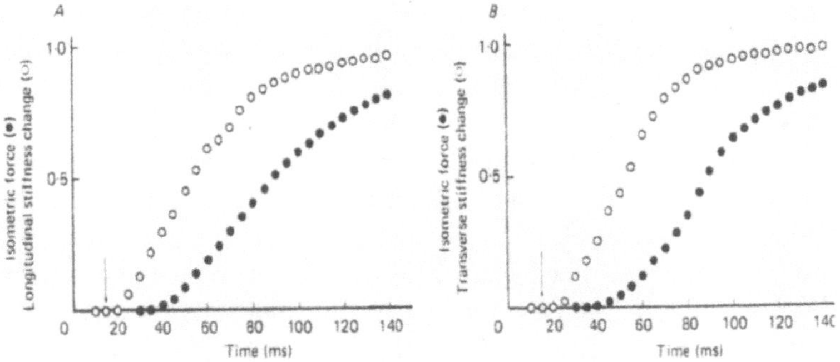

Figure 87 Muscle Stiffness (open circles) and isometric force (filled circles) changes, expressed relative to the maximum values attained. (A) Longitudinal stiffness and force. (B) Transverse stiffness and force. Note that both stiffness changes precede force changes. From Hatta et al. (1988).

In an extensive overview on the molecular mechanism of muscle contraction, Oplatka (1997) regarded our ultrasonic wave experiments as evidence for the electrostatic actin–myosin linkages with highly anisotropic nature, and discussed the possible electrostatic mechanism of the actin–myosin interaction in contracting muscle.

8.7.2 Other Evidence for the Electrostatic Actin-Myosin Interaction

The importance of electrostatic interactions between myosin heads and actin filaments has also been suggested by biochemical studies on acto-S1 in solution (Trybus and Taylor, 1980; Geeves and Lehrer, 1994; Lehrer, 1994), crystallographic studies of myosin head (Holmes, 1995; Rayment et al., 1993).

Meanwhile, Elliott and his coworkers have obtained evidence for the electrostatic mechanism of cyclic actin–myosin interaction in contracting muscle from their extensive work on the charge distribution of myofilaments during muscle contraction, which is triggered by Ca^{2+} release from sarcoplasmic reticulum and is maintained by hydrolysis of ATP (Elliott, 2007). Very sadly, Dr. Elliott passed away several years ago. However, his coworker Dr. Bartels recently wrote an exciting overview (Bartels, 2016) on the possible electrostatic mechanism of actin–myosin interaction producing muscle contraction. In this overview, she emphasizes changes in the ionic environment around 3D myofilament lattice, consisting of charged proteins, as the cause of muscle contraction based on experimental results obtained by Elliott and his coworkers.

Bartels and Elliott (1985) measured Donnan potentials in skinned muscle fibers, and showed that in rigor fibers, the fixed charge concentration in the A- band was higher than that in the I-band, while the fixed charge concentration was equal between the A- and the I-bands. Using concentrated gels of myosin and myosin rod, Bartels et al. (1993) have reported that the fixed electric charge in the A-band is localized on myosin filaments and that the fixed charge on myosin changes markedly in the presence of ATP. The ATP-induced myosin fixed charge is independent of actin filaments. The above myosin fixed charge changes would read to myosin conformational changes associated with myosin head power stroke. Bartels (2017) regards this idea to be consistent with our EC experiments that in the absence of actin filaments, myosin heads exert recovery stroke to take position ready for power stroke producing contraction.

If the electrostatic mechanism in actin–myosin interaction is correct, the terms strong and weak myosin head–actin filament binding are replaced with new terms: strong and weak electrostatic attractive forces at the myosin head–actin filament interface.

The twofold increase of force generated by individual myosin heads at low ionic strength (Sugi et al., 2013), which is supported by EC experiments (Sugi et al., 2015) can also be accounted for most readily by the decreased screening effect of ions to reducing strength of electrostatic myosin head–actin interaction.

In water solutions, the myosin head–actin binding is weak in A-M-ADP-Pi and strong in A-M-ADP and A-M (Bagshaw, 1993). On this basis, Pi should detach from A-M-ADP-Pi as soon as M attaches to A in the form of M-ADP-Pi to make A-M binding strong enough to exert power stroke producing myofilament sliding. Recently, however, it has been reported that myosin head power stroke starts before Pi release from A-M-ADP-Pi (Muretta et al., 2016), suggesting that, contrary to the general view (Bagshaw, 1993), myosin heads in the form of M-ADP-Pi can form not only weak binding but also strong binding with actin. This issue seems to be related to our finding that myosin heads, with anti-LD or anti-S-2 antibodies attached, exhibit high ATPase activity without producing no appreciable internal resistance against muscle shortening (Sugi et al., 1992, 2014).

Chapter 9

The Fenn Effect: A Fundamental Mystery in Muscle

In this chapter, I will explain the Fenn Effect, a most fundamental mystery in the research field of muscle contraction, since its discovery is a typical example that a dogma can be broken by an unexpected discovery. Although the Fenn effect has already been explained qualitatively by A. F. Huxley's contraction model (A. F. Huxley, 1957), it is concerned with a most important mystery as to how a muscle can be aware of the amount of load imposed on it and can regulate its energy output.

9.1 Historical Discovery of the Fenn Effect

In the early 20th century, A. V. Hill explained the mechanical response of contracting muscle by a viscoelastic model, consisting of damped and undamped springs connected in series (Fig. 6B). This model was widely believed by muscle physiologists over many years. On the basis of the viscoelastic model, the total energy output (heat + work) for a given amount of shortening should be constant irrespective of experimental conditions.

Although Hill developed a thermopile device to measure heat production in contracting muscle, it was not easy to prove

Mysteries in Muscle Contraction: Evidence against Current Dogmas
Haruo Sugi

ISBN 978-981-4774-48-2 (Hardcover), 978-1-315-10191-0 (eBook)
www.panstanford.com

the above prediction, because of complex nature of heat produced in muscle; according to the current knowledge, muscle heat production results from at least three components: (1) ATP hydrolysis to supply energy to contracting muscle, (2) Ca^{2+}-cycling between sarcoplasmic reticulum and myoplasm, and (3) thermoelastic heat associated with the work done on contracting muscle by experimental apparatus.

In 1923, O. W. Fenn, a young investigator who came from the USA to Hill's laboratory, carried out a historical experiments. He used an inertial lever system shown in Fig. 88. On electrical stimulation, the muscle first shortens by moving the inertial lever, and then goes slack after its contractile activity is over by the continuing movement of the inertial lever.

After the complete cessation of mechanical activity, the muscle is passively lengthened to its initial length by a light weight *W* attached to the isotonic lever. With this device, Fenn could record the heat produced by muscle shortening without complication arising from thermoelastic heat produced during subsequent passive muscle lengthening. Thus, Fenn showed that contracting muscle can regulate its energy output depending on experimental conditions.

Thus, Hill's viscoelastic model, which was a most prominent dogma in the field of muscle physiology in the early 20th century, was broken by one of his pupils. In his book published in 1970 (Hill, 1970), Hill wrote, "Fenn's conclusions were obviously the death warrant of the visco-elastic theory." Professor Katz (Nobel Prize winner in 1970), who stayed at Hill's laboratory at that time, told me that Fenn's experiments were performed on the basement floor of Hill's private house; at that time, investigators sometimes carried out experiments on the basement floor, where temperature was relatively constant. Katz also told me how Hill was unhappy with Fenn's results.

A most significant meaning of the Fenn effect is that biological systems are so complicated, containing a variety of self-regulatory systems, against the general view at that time that biological systems were simple and could be simulated by combination of physical elements. Of course, Fenn's conclusion is now a common sense to every investigator in the field of life science.

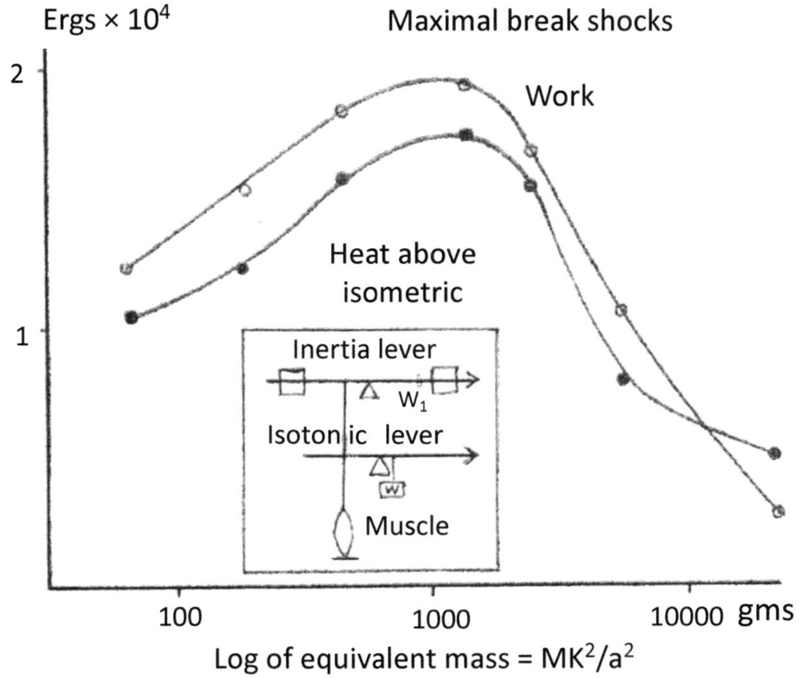

Figure 88 Figure reproduced from Fenn's original paper (Fenn, 1923), showing the amount of energy output (heat + work) from contracting muscle is not constant. As shown in the inset, muscle first contracts to move the inertia lever, so that it goes slack by continuing movement of inertial lever. After complete cessation of contractile activity, muscle is lengthened to its original length by a light weight *W* attached to isotonic lever.

9.2 Hill's Force–Velocity Equation

One of the basic characteristics of the contracting muscle is that when the load on isometrically contracting muscle is suddenly reduced from P_0 (the maximum isometric force) to $P < P_0$, the muscle shortens with a constant velocity (Fig. 6D); namely, contracting muscle shortens with a constant velocity (V) under a constant load (=force, P).

Hill (1938) studied the relation between V and P on contracting frog skeletal muscle and reported that the shape of the P–V relation was a rectangular hyperbola, so that

$$(P + a)\ (V + b) = \text{constant} = (P_0 + a)b, \text{ or}$$

$$(P + A)\ V = b\ (P_0 - P),$$

where a and b are constants (Fig. 89).

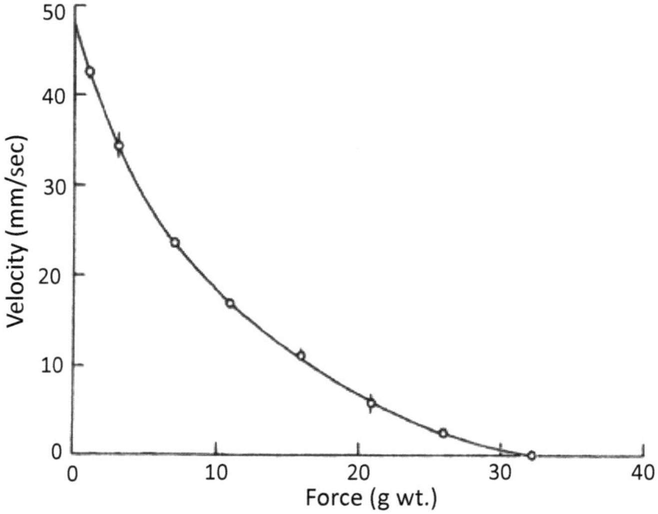

Figure 89 Force–velocity (P–V) relation of frog skeletal muscle. The curve fits part of rectangular hyperbola, so that the relation between shortening velocity V and force (= load) P is expressed as $(a + P)V = b\ (P_0 - P)$. From Hill (1938).

Hill claimed that the constant a was equal to the heat constant a, i.e., the amount of heat produced per unit distance of muscle shortening. The result that the two constants obtained from mechanical and thermal measurements are identical to each other attracted investigators' attention, and the above equation was regarded as indicating the fundamental property of contracting muscle to regulate the energy output (heat + work) depending on the load imposed on it. As a result, this equation was highly evaluated by muscle investigators over many years. This dogma was broken by Hill himself by showing that the two constants were not identical (Hill, 1964a).

Roughly speaking, however, the values aV and PV can be approximately regarded as heat flux and work flux in contracting muscle, respectively, and the constant b is sometimes called the constant of energy liberation. Of course, such a consideration gives no information about the mechanism, with which contracting muscle regulates its energy output depending on the load. Despite the fact that the Hill equation is now a mere empirical one, many investigators in the field of exercise physiology tend to discuss the significance of their results on the basis of the Hill equation, especially with respect to the value of a/P_0, which determines the curvature of hyperbolic P–V curve, representing muscle mechanical efficiency expressed as (work)/(heat + work).

Meanwhile, in the field of basic muscle physiology, the P–V relation obtained from the whole muscle, consisting of many different types of muscle fibers, nerve fibers, blood vessels, and connective tissues, is no longer of fundamental importance in the light of experiments with single muscle fibers, as will be explained in the following sections.

9.3 Force-Velocity Relation Obtained from Experiments with Single Muscle Fibers

Basic contraction characteristics of contracting muscle have been obtained mostly from the whole frog skeletal muscle. In addition to its complex composition, whole muscle deteriorates rapidly partly due to the overstimulation of component muscle fibers in contact with stimulating electrodes and partly due to insufficient oxygen supply to muscle fibers distant from the muscle surface. These problems are removed by using single intact or skinned muscle fibers.

9.3.1 Advantages of Single Muscle Fibers over Whole Muscles

Although the isolation of single fibers from the muscle is technically difficult, the use of single fibers has the following advantages over the use of whole muscles:

(1) A single muscle fiber is a structural and functional unit of muscle at the cellular level and therefore free from all complications arising from whole muscle.

(2) Single muscle fibers, isolated successfully from whole muscle, survive for many hours to yield many reproducible results.

(3) Striation pattern (sarcomere spacing) of muscle fibers can be easily observed under a light microscope, so that changes in the sarcomere length can be recorded during experiments.

Due to these advantages, experiments with single muscle fibers provide information to obtain insights into how contracting muscle can regulate its energy output.

On the other hand, it is difficult to change environments around myofilaments, since the surface membrane of intact muscle fibers acts as a barrier against diffusion of physiologically active substances such as ATP. To overcome this problem, skinned fibers, from which the surface membrane is removed either mechanically or chemically, have been developed.

Preparation of single fibers from bundles of skinned fibers is relatively easy, since a short muscle fiber segment (a few to several millimeters in length) is dissected from skinned fiber bundles (for technical reasons, rabbit psoas muscle is normally used) by cutting both ends. The single skinned fiber preparation is placed in the experimental setup by gluing both ends to transducers. As a result, skinned fiber preparation inevitably includes compliant and fragile cut region at both ends.

Despite these problems, skinned fiber preparation has been widely used because of its great advantage to be able to change the composition of experimental solutions around myofilaments. In addition, it is possible to record structural changes of muscle proteins constituting myofilaments, using various probes. It is also possible to measure ATPase activity at rest and during contraction, by measuring the amount ADP or Pi diffusing out of the fiber with NADH fluorescence or Pi-sensitive protein.

9.3.2 Force–Velocity Relation Obtained from Single Intact and Skinned Muscle Fibers

Edman (1988) reported that the *P*–*V* relation obtained from single intact muscle fibers was double-hyperbolic in shape, and deviated from the hyperbola at loads $\geq 0.7\ P_0$ (Fig. 90), so that a distinct hump appears in the *P*–*V* curve in the high force region. This indicates that the kinetics of actin–myosin interaction is different between low and high force regions. The double hyperbolic shape of the *P*–*V* relation is also observed by us in both intact and skinned single muscle fibers (Iwamoto et al., 1990; Sugi et al., 1992, 2013, 2014) (Figs. 49, 53, 56, and 61).

It seems possible that the different kinetics of actin–myosin interaction observed on single intact and skinned muscle fibers corresponds to the two different modes of the myosin head power stroke depending on the amount of force generated by individual myosin heads (Fig. 65) found in our EC experiments.

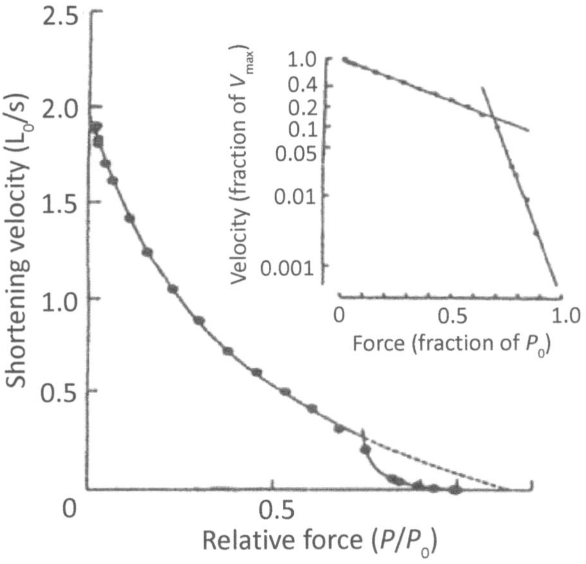

Figure 90 Double-hyperbolic P–V relation obtained from single intact frog muscle fibers. Inset shows semi-logarithmic plot of the same P–V data. From Edman (1988).

9.3.3 Isotonic Velocity Transients Preceding Constant Velocity Shortening in Single Intact Muscle Fibers

Experiments with the whole muscle are inadequate in recording muscle shortening at its early phase due to mechanical vibrations of massive experimental apparatus. Using single intact muscle fibers, Civan and Podolsky (1966) examined the early phase of muscle fiber shortening following quick decrease in load from P_0 to $P < P_0$, and found that non-steady oscillatory changes in shortening velocity preceded subsequent constant velocity shortening (Fig. 91). In general, the initial shortening velocity is faster than that of steady shortening and then slows down and increases again to eventually reach constant velocity.

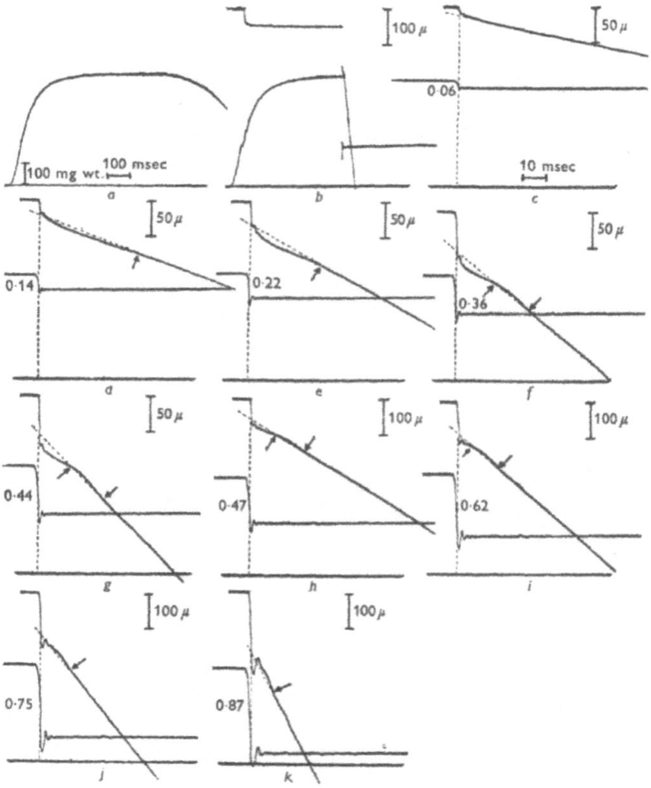

Figure 91 Non-steady oscillatory motion (upper traces) preceding constant velocity shortening following quick changes in load from P_0 to $P < P_0$ (lower traces). The value of P relative to P_0 is indicated in each record. Dotted line in each record indicates back extrapolation of constant velocity shortening to show oscillatory motion around each broken line. Civan and Podolsky (1966).

The transient oscillatory changes in shortening velocity preceding the following steady shortening gives me an impression that in response to a quick decrease in load, some feedback mechanism is operating to adapt new mechanical condition to decide steady shortening velocity, as if the fiber takes the new condition into consideration. The possible feedback mechanism may correspond to the mechanism underlying the Fenn effect.

We also examined the transient changes in shortening velocity not only following quick decreases in load, but also quick increases in load from P_0 to $P > P_0$ (Sugi and Tsuchiya, 1981a). As shown in Fig. 92, the transient changes in shortening velocity was similar to that reported by Civan and Podolsky (1966) as can be seen in left records (Fig. 92B), while oscillatory changes in lengthening velocity under a constant load $P > P_0$ was only seen with $P \leq 1.2\ P_0$ as can be seen in right records (Fig. 92B). In general, the fibers did not exhibit constant velocity lengthening; the velocity of lengthening was maximum at the beginning of lengthening and decreased as the lengthening went on. A most remarkable feature of fiber lengthening was that following rapid increases in load from P_0 to 1.17~1.07 P_0, the fibers showed alternate lengthening and shortening as shown in the lower records of Fig. 92B.

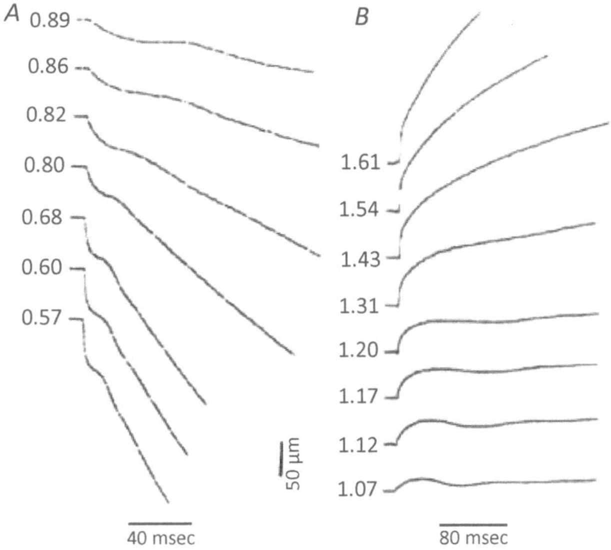

Figure 92 Transient changes in velocity of fiber shortening (A, left) and those in velocity of fiber lengthening (B, right) after rapid changes in load from P_0 to P. Values of P relative to P_0 are shown alongside each record. From Sugi and Tsuchiya (1981a).

The most remarkable alternate lengthening and shortening was observed when the load on the fibers was first reduced from P_0 to $P < P_0$ and then increased slightly from P after a period of fiber shortening (Fig. 93). Similar alternate lengthening and shortening has also been observed by Armstrong et al. (1966) and Edman and Curtin (2001), who showed that the oscillatory length changes took place uniformly along the fiber length, indicating that this phenomenon occurs in every sarcomere. In my view, the mechanism underlying the Fenn effect tends to be somewhat disturbed if small rapid changes in load around P_0 are applied at short intervals. Much more experimental work is necessary to reach full understanding of the mechanism, with which the contracting muscle regulates its energy output depending on mechanical conditions.

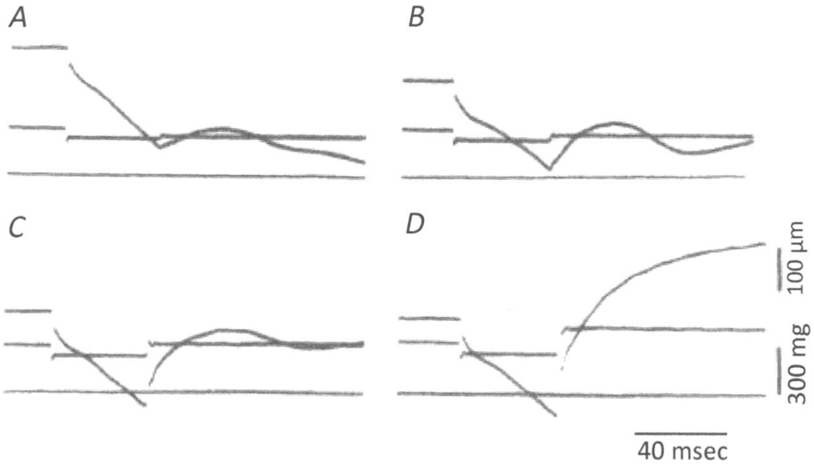

Figure 93 Marked alternate lengthening and shortening when the load on the fiber was first reduced, and then increased slightly after a period of fiber shortening. Upper traces show changes in fiber length, while middle and lower traces show changes in load (= force) starting from P_0 and zero force baseline, respectively. From Sugi and Tsuchiya (1981a).

9.4 The Fenn Effect Is Observed in Both Myofibrils and in vitro Motility Assay System

Here, I return to the Fenn effect that the energy output (heat + work) from the contracting muscle is maximum at moderate external load, under which it shortens with a moderate velocity (Fig. 88).

Experiments performed in our laboratory have shown that the Fenn Effect can be observed in laser flash-activated muscle myofibrils and also in an in vitro motility assay system, in which a myosin-coated microneedle moves along actin cables, indicating that the Fenn effect is a basic property of a functional unit in muscle.

9.4.1 Experiments on Muscle Myofibrils

We attempted to measure how ATPase activity associated with actin–myosin interaction is influenced by external loads by using a fluorescent nucleotide analog, Cy3-EDA-ATP. For this purpose, actomyosin solution is not suitable, because 3D myofilament lattice structure is lost. Single skinned muscle fibers are also not suitable, because of problems of diffusion of nucleotide analogs and background noise. For these reasons, muscle myofibrils are suitable for this purpose. In addition, it is necessary to use a confocal microscope to reduce background noise for recording fluorescence intensity changes in myofibrils. As shown in Fig. 94A, Cy3-EDA-ATP attaches to myosin heads in muscle myofibrils, as indicated by fluorescence images of a myofibril, which exhibits fluorescence in the A-band where myosin filaments are located. Both ends of the myofibril are held between two glass microneedles connected to a piezoelectric actuator (Fig. 94B).

In the absence of Ca^{2+}, Cy3-EDA-ATP is hydrolyzed by myosin heads but still attached to them in the form of Cy3-EDA-M-ADP-Pi. On simultaneous application of Ca^{2+} and ATP released from caged ATP following laser flash photolysis, myosin heads bind with the actin filament to form A-M-ADP-Pi, exert power stroke associated with release of Pi, and then detach from myosin heads in the form of Cy3-EDA-ADP (Cy3-EDA-nucleotide) to be replaced by non-fluorescent ATP. Consequently, it is possible to measure the rate of ATP hydrolysis of myofibril by recording the decrease in fluorescence of Cy3-EDA-ATP.

Figure 95A presents typical successive video images of the displacement of Cy3-EDA-nucleotide bound to myosin heads in the myofibril following laser flash photolysis of caged ATP during isometric contraction. Figure 77B shows the fluorescence decay curve giving a rate constant of 0.56 s^{-1}.

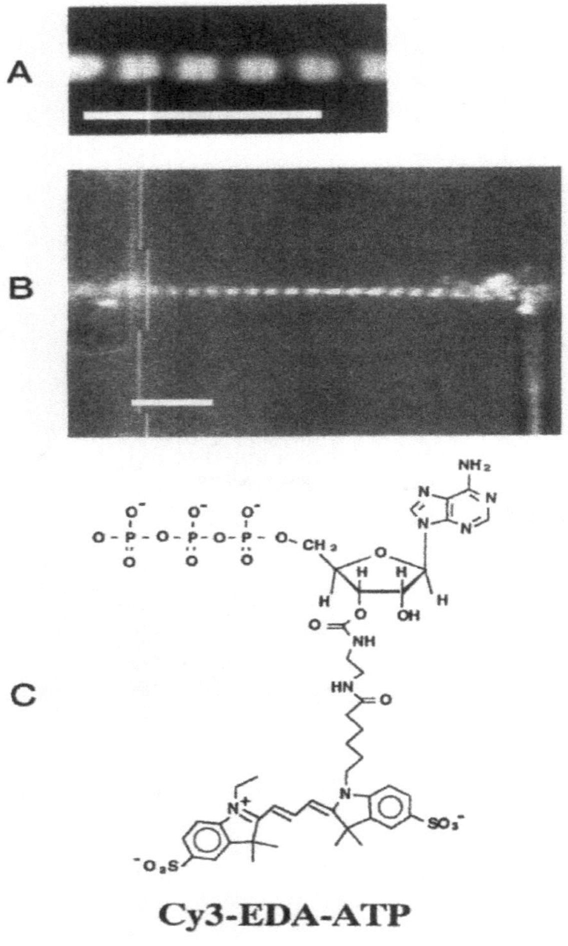

Figure 94 (A) Fluorescent images of single rabbit skeletal muscle myofibril. (B) Video image of a fluorescent myofibril with both ends attached to a pair of glass microneedles. Scale bar, 10 μm. (C) Structure of Cy3-EDA-ATP. From Chaen et al. (1997).

By performing experiments not only during isometric contraction, but also during isovelocity shortening of myofibrils, we obtained the result shown in Fig. 96. The rate of displacement of Cy3-EDA-nucleotide was maximum when the myofibril was made to shorten with a moderate velocity, and minimum when the myofibril was kept isometric. This result implies that the Fenn effect can be observed at the level of myofibrils.

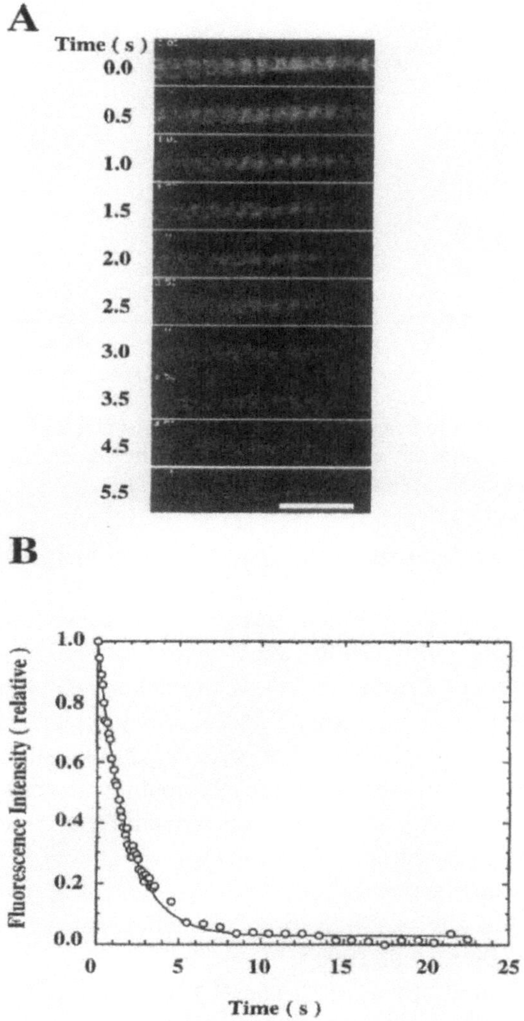

Figure 95 (A) Displacement of Cy3-EDA-nucleotide bound to myosin heads in a myofibril following laser flash photolysis of caged ATP under isometric condition. Scale bar, 10 μm. (B) Quantitative analysis of the displacement of Cy3-EDA-nucleotide. The data points are fitted to a single exponential, giving a rate constant of 0.56 s^{-1}. From Chaen et al. (1997).

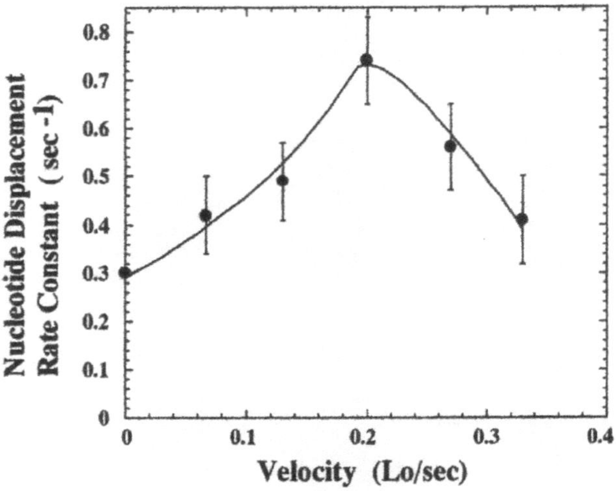

Figure 96 Relation between the Cy3-EDA-nucleotide displacement rate constant and the velocity of myofibril shortening. Each data point represents mean ± SD (n = 5). Chaen et al. (1997).

9.4.2 Experiments on in vitro Motility Assay System

Using an in vitro movement assay system consisting of myosin-coated glass microneedle and actin cables in giant algal cells (Fig. 21), we studied the relation between the amount of iontophoretically applied ATP and the amount of work done by the ATP-induced sliding of myosin coated-needle along actin cables. To eliminate the effect of ATP remaining in the experimental solution, hexokinase and D-glucose were added to the experimental solution to serve as an ATP scavenger. As shown in the inset of Fig. 97A, the needle stopped moving soon after termination of ATP current pulse, indicating the effectiveness of the ATP scavenger (Oiwa et al., 1991).

The amount of work (W) done by myosin heads on the needle during the course of ATP-induced movement from position x_1 to position x_2 on actin cables is

$$W = K\{(x_2)^2 - (x_2)^2\}/2,$$

where K is the elastic coefficient of the needle. As can be seen in Fig. 97B, The work versus baseline force relation is bell-shaped. If the baseline force at which the needle starts moving is regarded as the amount of afterload in the contracting muscle, the curve may be analogous with the bell-shaped relation between energy output and load relation in the contracting muscle (Fig. 88). Since the number of myosin heads producing needle movement may be about 10, these results suggest that the Fenn effect in the contracting muscle results from the characteristics of individual myosin heads.

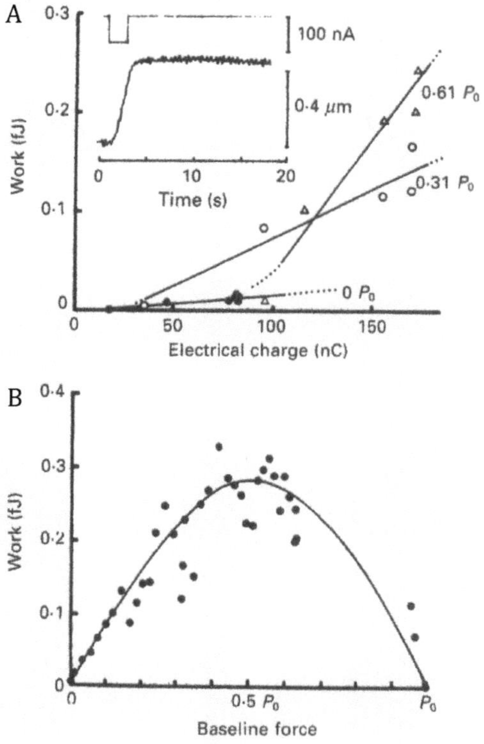

Figure 97 ATP-induced movement of a myosin-coated needle along actin cables. (A) Dependence of the amount of work done by ATP-induced microneedle sliding along actin cables (in fJ) and the amount of electrical charge passed through ATP-containing electrode (in nC). Needle movements were induced from three different baseline forces; zero (filled circles), 0.31 P_0 (open circles), and 0.61 P_0 (open triangles). Inset, typical needle movement without slow movement after termination of current to ATP electrode, indicating effectiveness of the ATP scavenger. (B) Relation between the amount of work done by needle movement (fJ) induced by ATP current (2s, 85 nA) and amount of baseline force (relative to P_0). Oiwa et al. (1991).

Chapter 10

Load-Dependent Mechanical Efficiency of Myosin Head Performance

Muscle is regarded as a kind of engine utilizing ATP as its fuel. In other words, muscle is a machine that converts the energy derived from ATP hydrolysis into mechanical work and heat. Although ATP was discovered in 1929, time course of ATP consumption by contracting muscle was not possible for many years. Consequently, the efficiency of the contracting whole skeletal muscle was estimated indirectly by Hill and coworkers by recording the work done and the heat produced during muscle contraction.

These classical experiments were followed by the experiments with single muscle fibers, in which the time course of ATP consumption can be recorded as a decrease in NADH coupled to ADP production (Bagshaw, 1993) in contracting muscle fibers. Although ATP consumption in contracting muscle can also be recorded by the method of nuclear magnetic resonance (NMR), the sensitivity of this method is not enough to perform single fiber experiments. In this section, I will describe and discuss our experiments on the mechanical efficiency of individual myosin heads within single skinned muscle fibers.

Mysteries in Muscle Contraction: Evidence against Current Dogmas
Haruo Sugi

ISBN 978-981-4774-48-2 (Hardcover), 978-1-315-10191-0 (eBook)
www.panstanford.com

10.1 Classical Experiments on Mechanical Efficiency of Whole Muscles

The ratio of work output rate (ΔW) to energy output rate ($\Delta W + \Delta H$) as measured is called mechanical efficiency. In whole frog skeletal muscle, the mechanical efficiency versus isotonic load relation, as determined during a period of isotonic shortening, is bell-shaped with a peak of ~0.45 at load ~0.5 P_0 (Fig. 98, Hill, 1964b). The mechanical efficiency was also obtained from the relation of total work output (W) to the total energy output ($W + H$), with a peak at 0.39 in frog skeletal muscle (Hill, 1939). As already explained, the complex nature of heat production in contracting muscle, together with the complex composition of whole muscle, makes the measurement of mechanical efficiency ambiguous.

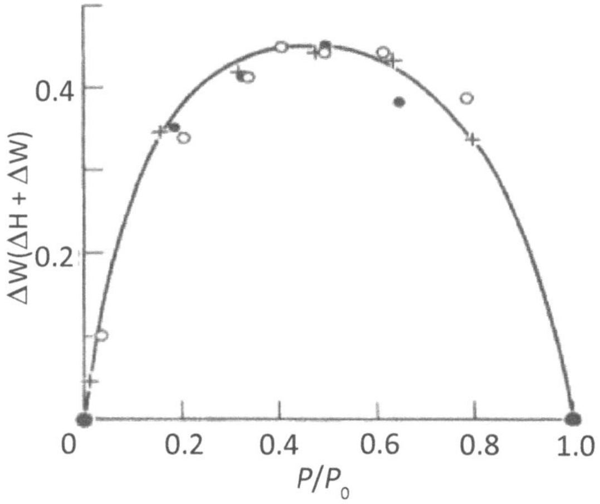

Figure 98 Bell-shaped mechanical efficiency versus load relation in whole frog skeletal muscle. From Hill (1964b).

It is now well established that ATP is the immediate source of energy for muscle contraction. The mechanical efficiency of muscle contraction is now estimated by measuring the work done by contracting muscle fibers and the amount of ATP hydrolyzed by measuring the amount of decrease in NADH coupled to ADP production.

The values of the maximum mechanical efficiency thus obtained, however, show a wide range of variation from 0.2 to 0.46 (He et al., 1997; Reggiani et al., 1997; Sun et al., 2001). It seems possible that the net maximum mechanical efficiency of myosin heads during their asynchronous activity is much smaller than the mechanical efficiency of individual myosin heads. We carried out the following experiments to make this point clear.

10.2 Measurement of Mechanical Efficiency of Individual Myosin Heads during Auxotonic Shortening

10.2.1 Experimental Design

We were interested in studying the load-dependent mechanical efficiency in individual myosin heads without complications arising from asynchronous myosin head movements producing power and recovery strokes. For this purpose, we designed the following experiments, based on the Lymn–Taylor scheme (Fig. 18) (Sugi et al., 1998). According to this scheme, myosin heads (M) in the relaxed muscle are in the state of M-ADP-Pi, and when skinned muscle fibers are activated with externally applied Ca^{2+}, they attach to actin filaments (A) to from A-M-ADP-Pi to exert power stroke almost synchronously associated with the following reaction:

A-M-ADP-Pi → A-M + ADP + Pi

If the ATP concentration in the experimental solution is previously made to be almost the same as that of myosin head concentration within a muscle fiber, almost all myosin heads in the fiber are expected to form M-ADP-Pi with long life times (>10 s). When the fiber is then rapidly activated by laser flash photolysis of caged Ca^{2+} (DM-nitrophen), almost all myosin heads are expected to exert power stroke synchronously and then take the form of A-M. In other words, almost all myosin heads in the fiber are expected to perform power stroke only once and then go into rigor state.

Based on the above idea, we first put single skinned muscle fibers in the experimental solution containing 220 μM ATP and 2.6 mM DM-nitrophen, and after 40 min exposed to air to be subjected to an intense flash (duration, 8 ns; wavelength, 350 nm; intensity, 20 mJ), so that the fiber was rapidly activated with Ca^{2+} released by the photolysis of DM-nitrophen. During the above procedures, the fiber did not exhibit any appreciable rigor force, indicating that the ATP concentration within the fiber did not decrease below the myosin head concentration (presumably ~200 μM). The fiber was connected to the lever of a servo motor at one end, while the other end was connected to a force transducer. Fiber shortening was sensed by a displacement transducer incorporated in the servo motor system.

10.2.2 Features of Auxotonic Shortening of Laser Flash-Activated Muscle Fibers

In the experiments to be described in this section, the force generated by flash-activated fiber was fed to the servo motor system in such a way that the resistance of servo motor arm against fiber shortening increased in proportion to the force developed. By this method, the fiber shortened under various auxotonic loads, i.e., against linear springs with various compliances.

Typical results are shown in Fig. 99. Interestingly, time courses of flash-induced auxotonic shortening against various auxotonic compliances, C_{auxso}, are identical when normalized with respect to the maximum extent of shortening.

We focused on the time course of decrease in the velocity of auxotonic shortening V. We expected sharp decrease in V with distance shortened as a result of increasing proportion of myosin heads forming rigor linkages with actin filaments after finishing their power stroke, since rigor linkages A-M provide considerable resistance against fiber shortening. Unexpectedly, the decrease of V with increasing distance of shortening was much slower than expected (Fig. 100); as indicated by broken lines, V decreased linearly with distance of shortening over a considerable time up to 2 s. In the above experiments, most myosin heads that finished their power stroke can no longer find the next ATP. Therefore, the slow decrease of shortening velocity over a considerable time strongly suggests that during fiber shortening, intense rigor linkages A-M are not readily formed between myosin heads taking their post-power stroke configuration and actin filaments. It seems possible that in the 3D myofilament-lattice structure, strong rigor A-M linkages are not present in contracting muscle. Therefore, the above features of auxotonic shortening also constitute evidence against the dogma (based on the Lymn–Taylor scheme) that, after the end of the power stroke, myosin heads pass through tight rigor A-M configuration.

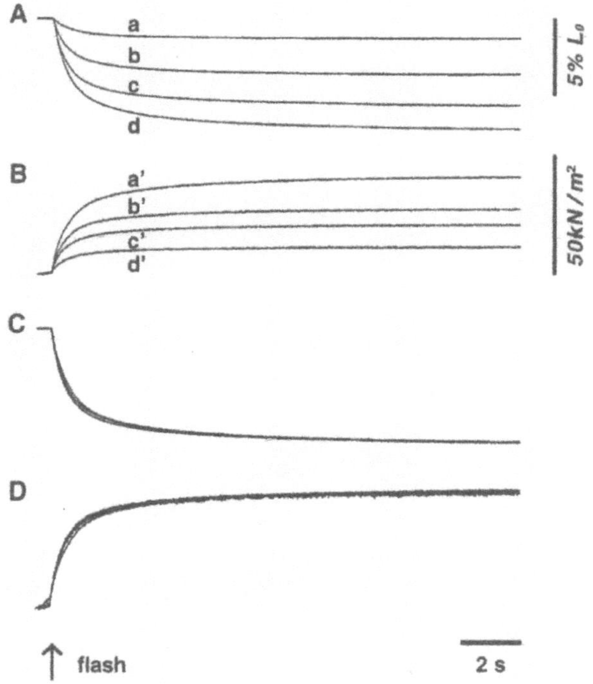

Figure 99 Fiber length (A) and force (B) changes during flash-induced fiber shortening against four different auxotonic loads. Length records a, b, c, and d correspond to force records a′, b′, c′, and d′, respectively. Normalized length and force changes with respect to the maximum values attained are shown in C and D, respectively. From Sugi et al. (1998).

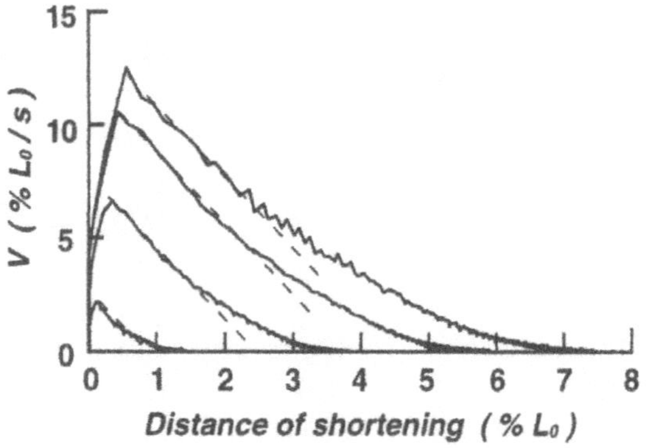

Figure 100 Relation between the velocity (V) and the distance of auxotonic fiber shortening against four different anxotonic loads. The records used are the same as those shown in Fig. 99. Broken lines indicate that after reaching a peak, V decreases linearly with increasing distance of shortening along the same slope irrespective of the amount of auxotonic load. From Sugi et al. (1998).

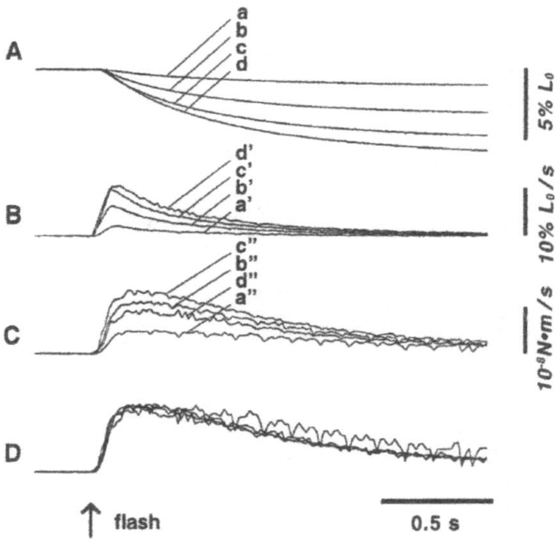

Figure 101 Changes in fiber length (A), shortening velocity (B), and power output (C). In the early phase of fiber shortening against four different auxotonic loads. Length records a, b, c, and d correspond to velocity records a′, b′, c′, and d′, and to power output records a″, b″, c″, and d″, respectively. Normalized power output records with respect to the peak values attained are shown in D. The records are the same as those shown in Fig. 99. From Sugi et al. (1998).

In Fig. 101 are shown the early phase of changes in fiber length, velocity of fiber shortening, and power output (shortening velocity times force), on a fast time base. As can be seen in Fig. 78C, the amount of work done by the fiber *W* (the area under the power output record) was maximum at a moderate auxotonic load. Since the value of *W* is zero in the isometric condition and in the practically unloader condition, the relation between the values of *W* and those of the auxotonic load, which can be replaced by the distance of fiber shortening for a given time, is bell-shaped as shown later.

10.2.3 Load-Dependent Mechanical Efficiency of Myosin Heads during Auxotonic Shortening

To estimate the amount of M-ADP-Pi utilized for mechanical work at a given time after the onset of flash-induced activation of the fiber, the fiber was first made to shorten under various auxotonic loads and after a period of auxotonic shortening (0.25~1 s) was subjected to a quick decrease in fiber length (quick release, complete in 1~2 ms) to drop the force just to zero. After the quick release, the fiber length was clamped, so that the isometric force developed at the decreased fiber length. The magnitude of the isometric force developed was taken as a measure of the amount of M-ADP-Pi, which remained in the fiber at the end of auxotonic shortening.

Typical results are shown in Fig. 102. It can be seen that the amplitude of force development after a quick release decreases with increasing distance of preceding auxotonic shortening, indicating that the amount of ATP (or more exactly the amount of M-ADP-Pi) utilized for fiber shortening increases with increasing distance of fiber shortening.

Figure 103 shows fiber length and force changes when the fiber was first made to contract isometrically and then subjected to a quick release (Fig. 103A), and when the fiber was first made to shorten under practically unloaded condition, and then shortening is interrupted by clamping fiber length (Fig. 103B), at 0.25 and 1 s after flash activation. In both cases, the magnitude of the isometric force developed P_r serves as a measure of the amount of M-ADP-Pi remaining in the fiber at 0.25 and 1 s after the onset of flash-activated isometric force generation or shortening under the practically unloaded condition, as illustrated in the inset in Fig. 103C. Based on this idea, the quantity $P_u = (P_o - P_r)$, represents the amount of ATP utilized for the flash-activated mechanical response.

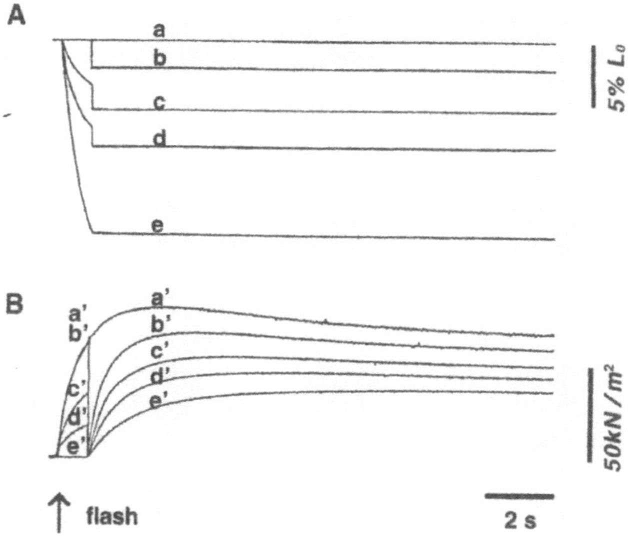

Figure 102 Fiber length (A) and force (B) changes when the fiber was first made to contract isometrically or shorten against various auxotonic loads for 1 s, and then subjected to a quick release to drop the force just to zero to develop isometric force at the decreased fiber length. Length records a, b, c, d, and e correspond to force records a′, b′, c′, and e′, respectively. The compliance of auxotonic load was zero (=isometric condition, records b-b′), and very large (practically unloaded condition, e-e′). Records a-a′ were obtained in the isometric condition, while records e-e′ were obtained in the unloaded condition. From Sugi et al. (1998).

In Fig. 104, the values of P_u, i.e., the amount of ATP utilized for mechanical response, and W, i.e., the amount of work done, expressed relative to the maximum value W_{max}, at 1 s after flash activation are plotted against the distance of fiber shortening at 1 s after flash activation, representing the compliance of auxotonic load. The value of P_u at a given time after flash activation was minimum in the isometric condition and increased with increasing compliance of the auxotonic load. The amount of work done was maximum at a moderate auxotonic load, and zero in the isometric and in the practically unloaded conditions, so that the work versus auxotonic load relation was bell-shaped.

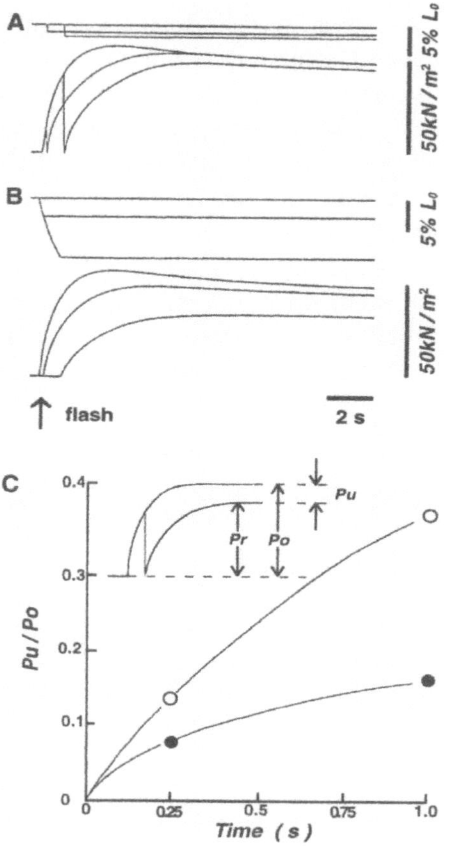

Figure 103 (A and B) Fiber length (upper traces) and force (lower traces) changes when the amount of ATP utilized for mechanical work P_u was estimated at 0.25 and 1 s after onset of flash activation in the isometric condition (A) and in the practically unloaded condition (B). In C, the values of P_u/P_o in the isometric (filled circles) and in the practically unloaded (open circles) conditions are plotted against time after flash activation. The records were obtained from the same fiber. Inset shows the measurement of P_u, representing the amount of ATP utilized, as $P_u = (P_o - P_r)$. From Sugi et al. (1998).

In the experiments described above, the most interesting feature of auxotonic fiber shortening was that both the fiber length and force changes were scaled in proportion to the maximum values attained as shown in Fig. 99C,D. As a result, the power output records were also scaled in proportion to the peak values attained in the early phase of fiber shortening (Fig. 101D).

At the beginning of fiber shortening, the amount of auxotonic load against fiber shortening is zero irrespective of the amount of auxotonic load. Consequently, the number of myosin heads in the state of M-ADP-Pi starting their power stroke is likely to be the same irrespective of the amount of auxotonic load. This may constitute a reason for the scaling of length and force records. Another reason for the scaling may be that tight rigor A-M linkages are not readily formed by myosin heads that finished their power stroke, at least until flash-induced fiber shortening comes to a complete stop; otherwise, the velocity of fiber shortening should fall abruptly due to increasing internal resistance against shortening arising from tight A-M linkages.

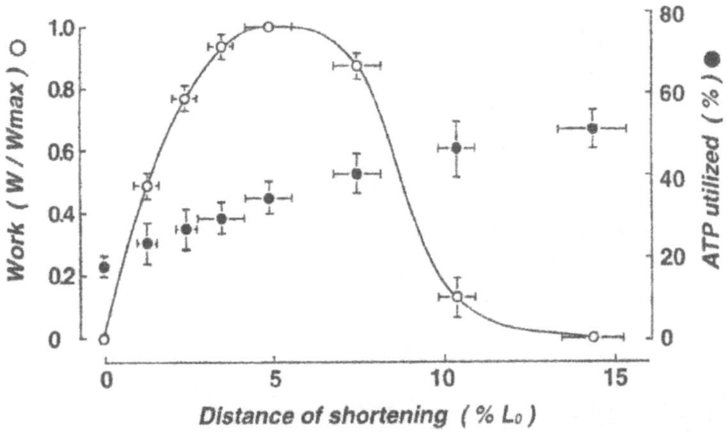

Figure 104 Dependence of the amount of ATP utilized for mechanical response, obtained from P_r/P_0 (filled circles) and the amount of work done, expressed relative to the maximum value W/W_{max} (open circles) on the distance of fiber shortening at 1 s after laser flash activation, representing the amount of auxotonic load. Vertical and horizontal bars indicate SEM. From Sugi et al. (1998).

The scaling of length and force changes found in this study indicates that myosin heads can sense the amount of auxotonic load in the early phase of shortening to determine their future mechanical response. Judging from the data in Fig. 100, myosin heads sense the amount of load and determine their future mechanical response until the shortening velocity reaches a maximum (until the fiber shortens by <1%). It seems possible that the load sensor resides in myosin head LD and myosin S-2, both of which are totally ignored in the swinging lever arm mechanism.

10.3 Measurement of Mechanical Efficiency of Individual Myosin Heads during Isotonic Shortening

10.3.1 Features of Power Output in the Early Phase of Isotonic Shortening

We also performed experiments to determine mechanical efficiency of individual myosin heads in single skinned muscle fibers during the afterloaded isotonic shortening, in which the fiber first contracted isometrically until the force generated by it reached the value equal to the load and then started shortening by lifting the load. During the fiber shortening, the load on the fiber was kept constant (isotonic shortening).

The design and the experimental apparatus used were the same as those in the auxotonic load experiments described in the previous section. Namely, almost all myosin heads were in the state of M-ADP-Pi, and on laser flash photolysis of caged Ca^{2+} (DM-nitrophen), the fiber was fully activated to shorten isotonically under various amounts of afterload.

Typical superimposed records of changes in fiber length (A) and force (B) are shown in Fig. 105. Except for the isometric condition (a, a′) and the unloaded shortening (f, f′), the fiber first generates isometric force, and when it reaches the level equal to the afterload, the fiber shortens isotonically under a constant load imposed by the servo motor system.

Due to the limited ATP concentration in experimental solution, almost all myosin heads within the fiber can hydrolyze ATP only once, so that the time course of isotonic shortening is not linear, but exponential, and after shortening for a limited distance the fiber stops shortening after complete exhaustion of ATP in the fiber.

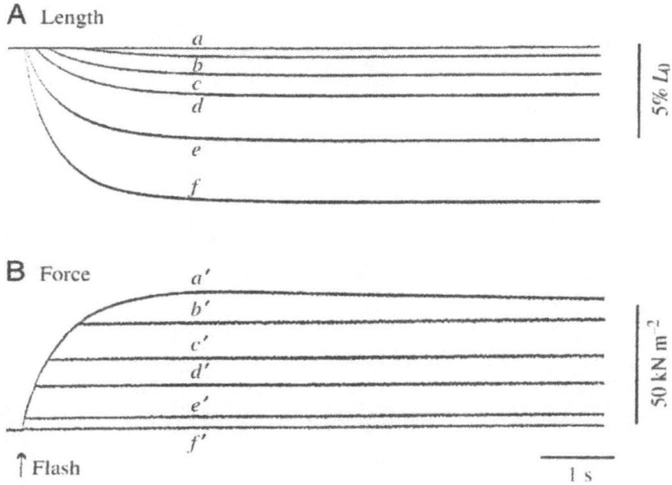

Figure 105 Laser flash–induced length (A) and force (B) changes in a single skinned muscle fiber. Length records a, b, c, d, e, and f correspond to force record a′, b′, c′, d′, e′, and f′, respectively. The amount of afterload was P_0 (isometric condition, a, a′), 0.78 P_0 (b, b′), 0.53 P_0 (c, c′), 0.35 P_0 (d, d′), 0.09 P_0 (e, e′), and zero (unloaded shortening, f, f′), respectively. From Sugi et al. (2003).

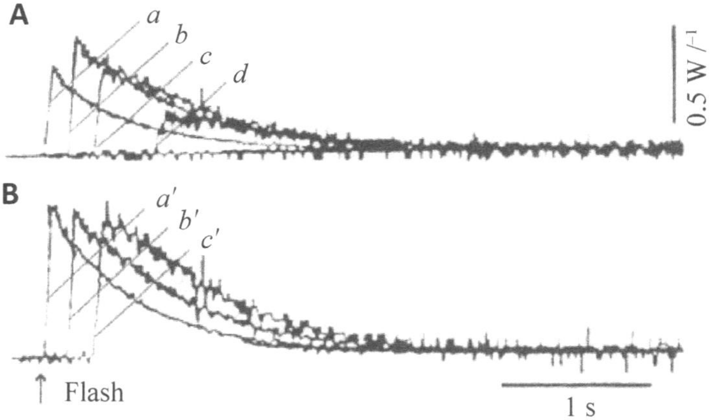

Figure 106 Time course of power output during flash-induced isotonic fiber shortening under four different afterloads (A), and the same records normalized relative to the peak values attained (B). The amount of afterload was 0.09 P_0 (a, a′), 0.35 P_0 (b, b′), 0.53 P_0 (c, c′), and 0.78 P_0 (d, d′), respectively. These data are obtained from the results shown in Fig. 105. From Sugi et al. (2003).

As shown in Fig. 106A, the power output reached a peak at the early phase of isotonic shortening and then decreased with time. The higher the initial peak, the larger the area under the power output trace, i.e., the amount of work *W* done by the fiber during isotonic shortening. The power output records were almost identical in shape, when normalized relative to the peak values attained, except for the loads close to P_0 (Fig. 106B). It is of interest that the distance of fiber shortening when the power output reached a maximum did not exceed 10 nm per half sarcomere, the value normally believed for the amplitude of myosin head power stroke (Huxley and Simmons, 1971). This strongly suggests that at the beginning of isotonic shortening, myosin heads in the form of M-ADP-Pi start their power stroke with amplitude of ~10 nm almost synchronously on binding with actin filament after the photolysis of caged-Ca, while sensing the amount of load to determine their future mechanical response. The possible location of the load sensor may reside in myosin head LD and myosin S-2, as already discussed.

10.3.2 Mechanical Efficiency of Myosin Heads during Afterloaded Isotonic Shortening

Figure 107 shows results of experiments, in which single skinned muscle fiber was first made to shorten under various amount of afterload, and subjected to quick releases to drop the force to zero at 1 s after flash activation. After quick releases, the fiber was allowed to develop isometric force at the decreased fiber length. Thus, the amount of ATP utilized for isotonic shortening, P_u was estimated as $P_u = (P_o - P_r)$, where P_o is the maximum isometric force, and P_r is the peak force developed after quick release (Fig. 103C, inset). The dependence of the amount of work done during isotonic shortening, *W*, and the amount of ATP utilized, P_u, on the amount of isotonic load, P_u is shown in Fig. 108. The value of P_u at $P = 0$ (unloaded shortening) was about three times larger than that at $P = P_o$ (isometric condition). The value of *W* was maximum at $P = \sim 0.4\ P_o$. The *W* versus *P* relation was also bell-shaped, since *W* is 0 at $P = 0$ and $P = P_o$.

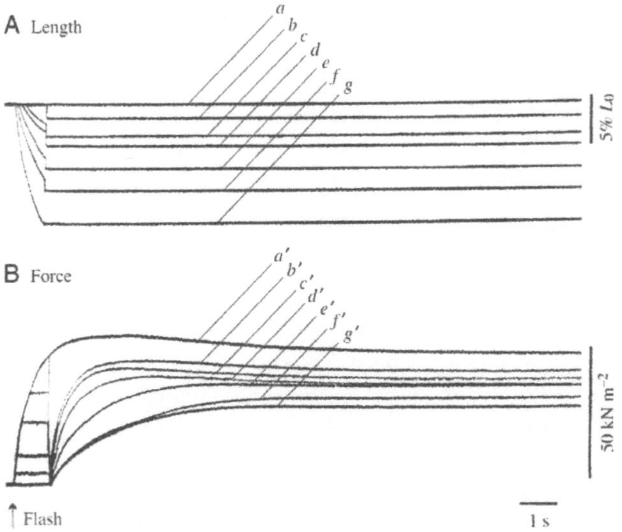

Figure 107 Fiber length (A) and force (B) changes, when a skinned fiber is first made to shorten isotonically under various afterload, and then subjected to quick releases to drop force to zero at 1 s after flash activation. After the interruption of isotonic shortening, the fiber was allowed to develop isometric force. Length records a to g correspond to force records a′ to g′, respectively. The amount of afterload was P_0 (isometric condition, a, a′), 0.63 P_0 (b, b′), 0.41 P_0 (c, c′), 0.20 P_0 (d, d′), 0.20 P_0 (e, e′), 0.09 P_0 (f, f′), 0 P_0 (unloaded condition, g, g′), respectively. Records a, a′ were obtained without quick release. From Sugi et al. (2003).

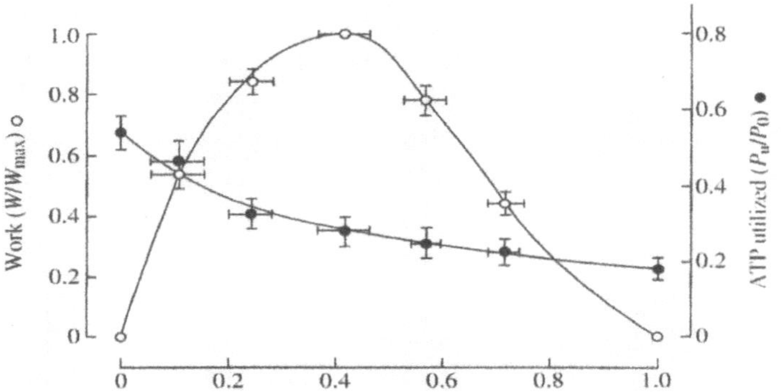

Figure 108 Dependence of the amount of ATP utilized for mechanical response (P_u, filled circles) and the amount of work done during isotonic shortening (W, open circles) at 1 s after flash activation. Vertical and horizontal bars indicate SEM. From Sugi et al. (2003).

10.3.3 Dependence of Mechanical Efficiency of Myosin Heads on Isotonic Load

The amount of ATP utilized for mechanical response P_u can be divided into two components: (1) ATP utilized for the development of isometric force preceding isotonic shortening, P_i, and (2) ATP utilized for the subsequent isotonic shortening, P_s. The values of P_i at various times after flash activation can be obtained from experiments, in which the fiber is first made to contract isometrically and is subjected to a quick release at various times after flash activation; the amount of P_i is determined from the magnitude of isometric force redevelopment after a quick release at any time after flash activation (Fig. 109).

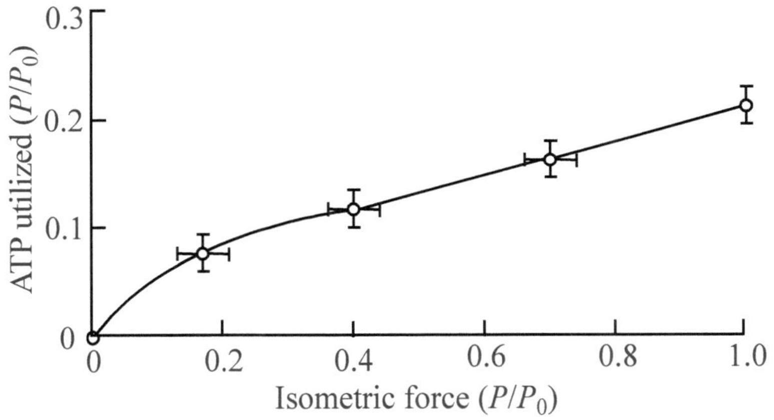

Figure 109 Relation between the amount of ATP utilized for isometric contraction P_u/P_o and the magnitude of isometric force P/P_o. From Sugi et al. (2003).

The results are shown in Fig. 110. By using the value of P_i (relative to P_u), it was possible to obtain P_s at any time after flash activation. The shape of mechanical efficiency versus isotonic load relation was somewhat skewed bell shape, being maximum at P = ~0.53 P_o and zero at P = 0 (unloaded shortening) and $P = P_o$ (isometric condition).

We further obtained the absolute value of mechanical efficiency in individual myosin heads by choosing appropriate values for fiber volume, myosin head concentration in the fiber, number of M-ADP-Pi immediately before flash activation, and energy released by ATP hydrolysis, and obtained the maximum efficiency of individual myosin head of 0.7. Since we used conservative values in calculating the efficiency, this value is apparently smaller than the actual value. This implies that the actual maximum mechanical efficiency of the individual myosin head is close to unity. In this connection, it is of interest that the mechanical efficiency of ATP-dependent rotary motion of F0-F1 ATPase in the mitochondrial membrane is reported to be close to unity (Kinosita et al., 2000).

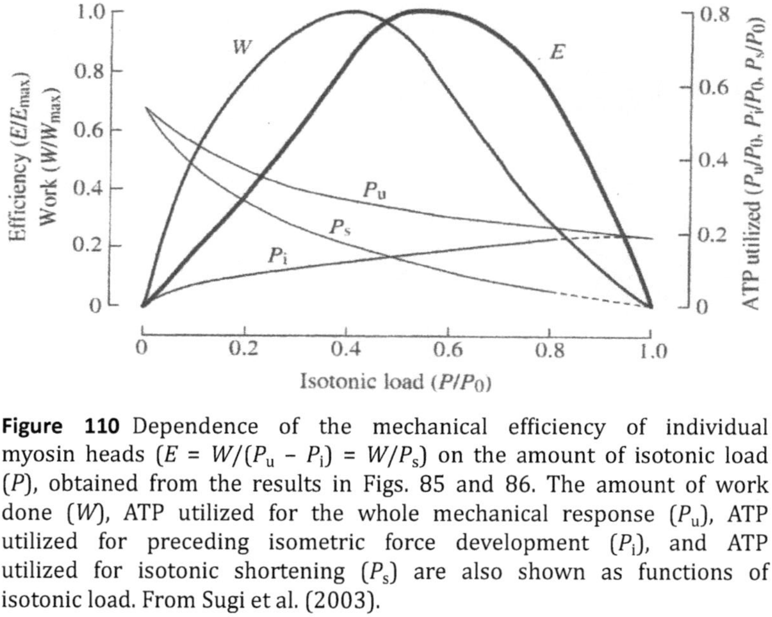

Figure 110 Dependence of the mechanical efficiency of individual myosin heads ($E = W/(P_u - P_i) = W/P_s$) on the amount of isotonic load (P), obtained from the results in Figs. 85 and 86. The amount of work done (W), ATP utilized for the whole mechanical response (P_u), ATP utilized for preceding isometric force development (P_i), and ATP utilized for isotonic shortening (P_s) are also shown as functions of isotonic load. From Sugi et al. (2003).

Chapter 11

Past History-Dependence of Myosin Head Performance

11.1 Force Enhancement after Slow Stretch and Force Deficit after Slow Release

In 1952, Abbott and Aubert showed that when contracting muscle generating steady isometric force is slowly stretched, the force rises during stretch and after the end of the stretch decays to a steady level higher than the normal isometric force at the stretched muscle length (Fig. 111, upper records). Conversely, if contracting muscle is slowly released, the force falls during release, and at the end of the release rises to a steady level lower than the normal isometric force at the released muscle length (Fig. 111, lower records). These phenomena are called force enhancement after slow stretch and force deficit after slow release. H. E. Huxley (1960) pointed out that these phenomena may arise from change in the performance of myosin heads when they are slowly displaced in either direction. The past history-dependence of myosin head performance cannot be readily explained as yet and also remains a mystery. In this chapter, I will first describe some interesting features of this phenomenon, mainly revealed by our experiments.

Mysteries in Muscle Contraction: Evidence against Current Dogmas
Haruo Sugi

ISBN 978-981-4774-48-2 (Hardcover), 978-1-315-10191-0 (eBook)
www.panstanford.com

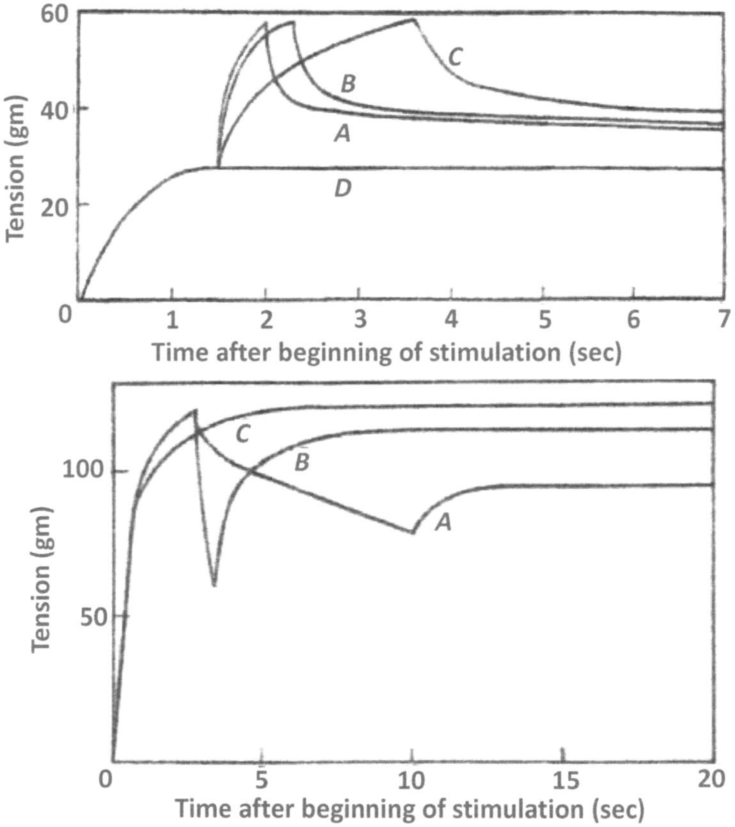

Figure 111 Force (tension) changes in contracting dogfish jaw muscle in response to slow stretches (upper records) and in response to slow releases (lower records). Note that the steady force levels attained after stretches or releases depend on the velocity of length changes applied. From Abbott and Aubert (1952).

11.2 Heat Production of Whole Muscle during Slow Stretch

In 1959, Hill and Howarth measured heat production of isometrically tetanized whole frog skeletal muscle during a slow stretch with the results illustrated in Fig. 112. During stretch, the force in the muscle continued to rise, while the rate of heat production increased markedly by stretch. Rather surprisingly, the sum of the heat produced (*h*) by the muscle and the work done on the muscle by the experimental apparatus (*Wap*) showed a time course indicated by record 3 in Fig. 93C, giving an impression that the chemical reactions associated with ATP hydrolysis are inhibited or even reversed during isometric force generation. Their results stimulated the interest of many investigators to examine whether ATP was actually synthesized in tetanized muscle during stretch or not with the results that reversal of chemical reactions never took place.

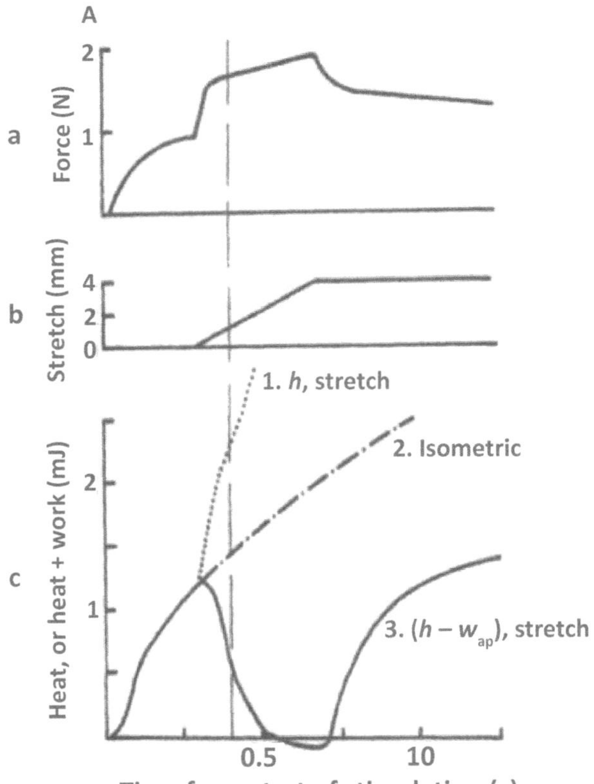

Figure 112 Effect of slow stretch on the force and the rate of heat production in isometrically tetanized muscle. Force and length changes of muscle are shown in (a) and (b), respectively. (c) Time course of heat or heat + work during stretch. Record 1, heat production during stretch. Record 2, heat production during an isometric tetanus. Record 3, heat (h) + work (W_{ap}) during stretch (Hill and Howarth, 1959).

In terms of the sliding filament mechanism, myosin heads interacting with actin filaments are forcibly displaced in the direction opposite to muscle shortening, and might slip along a number of myosin-binding sites on actin filaments without hydrolyzing ATP, as if they are locked-on not to perform normal cyclic interaction with actin filaments (H. E. Huxley, 1960). It has been, however, pointed out that the experiments of Hill and Howarth (1959) had a number of shortcomings, including non-uniform length changes along the muscle length (Woledge et al., 1985).

11.3 Enhancement of Mechanical Performance in Stretched Muscle Fibers: Possible Storage of Energy by Stretch

Using single muscle fibers, Edman et al. (1978) have shown that the mechanical performance of than fiber is enhanced by slow stretch, so that the force–velocity relation shifted toward higher force values, indicating that the work done on the fiber is stored, at least in part, in some sarcomere structures.

Sugi and Tsuchiya (1981b) have shown that the enhanced mechanical performance is also observed in muscle fibers during slow isotonic lengthening. Figure 113 presents a typical result of experiments, in which an isometrically tetanized fiber was first subjected to small load steps from P_0 to 1.2 P_0, and then again subjected to load steps from 1.2 P_0 to 1.1 P_0 at various times after the preceding load steps. It can be seen that the fiber shortens against the load of 1.1 P_0; in records c–e, the fiber starts shortening immediately after the load step, while in records a–c, some period is required for the fiber to start shortening after the load step. In Fig. 114, the isometrically tetanized fiber is first subjected to load steps from P_0 to 1.3–1.4 P_0 and made to shorten against various isotonic loads after a period (45 ms) of lengthening. The fiber shortens against the load close to P_0 (Fig. 114B), indicating clearly the enhanced mechanical performance.

Interestingly, the fiber starts shortening with steady velocities without exhibiting the isotonic velocity transients shown in Figs. 91–93. Since the isotonic velocity transients can be accounted for in terms of redistribution of myosin heads attached to actin filaments, the absence of the isotonic velocity transients in stretched or lengthened muscle fibers seems to suggest that the energy responsible for the enhanced mechanical performance may be stored in some structure other than the myosin head–actin filament connections. This point will be discussed later in connection with our X-ray diffraction studies.

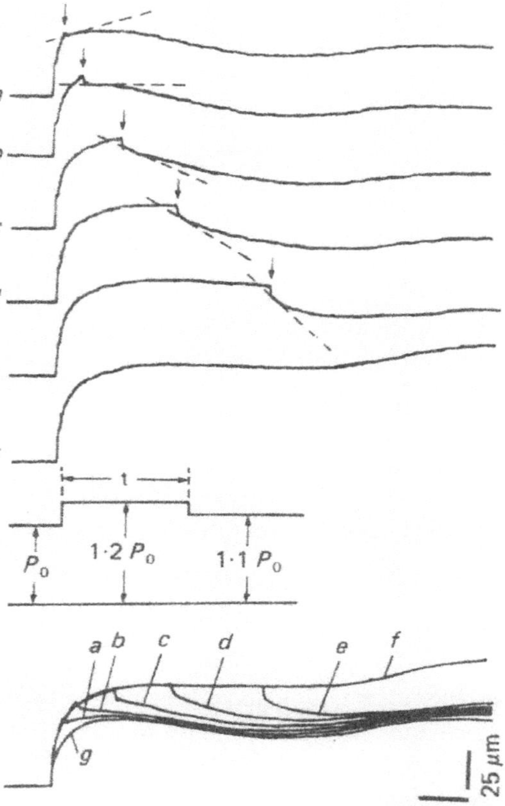

Figure 113 Length changes of a tetanized muscle fiber. The fiber was first tetanized isometrically, and then subjected to quick increase in load from P_0 to 1.2 P_0. At various times after the beginning of oscillatory isotonic lengthening under a load of 1.2 P_0, the fiber was further subjected to a quick decrease in load from 1.2 to 1.1 P_0 (records a–e). The second load step was not applied in record f. Superimposed tracings of records a-f, together with the length change to load step from P_0 to 1.1 P_0 are shown at the bottom (Sugi and Tsuchiya, 1981b).

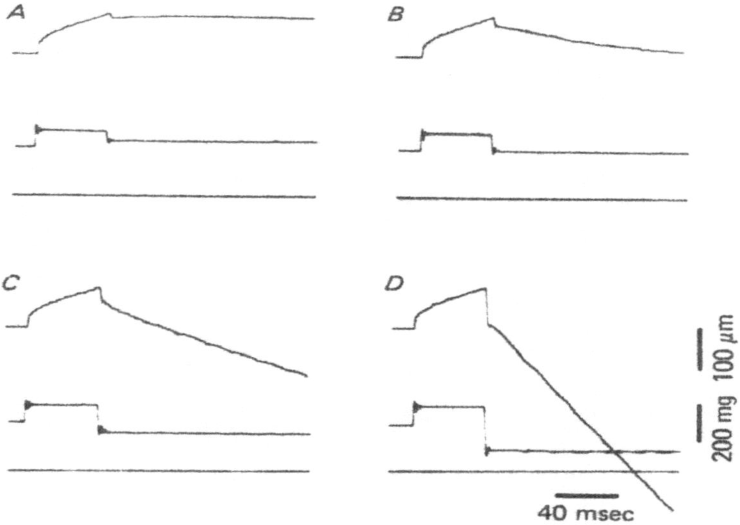

Figure 114 Changes in fiber length (upper traces) and force (=load) in a tetanized fiber when the load was quickly decreased from 1.3–1.4 P_0 to various values after a period (45 ms) of isotonic lengthening. Note that, in B the fiber shortens against the load close to P_0, and that the fiber starts shortening with steady velocities without exhibiting isotonic velocity transients shown in Figs. 72–74. From Sugi and Tsuchiya (1981b).

11.4 Measurement of Muscle Fiber Stiffness in Single Tetanized Muscle Fibers before, during and after Slow Stretches

To obtain further information about the mystery of enhancement of isometric force after slow stretches, we first confirmed that during applied slow stretch, tetanized muscle fibers were lengthened uniformly along their entire length, by putting carbon particles at various regions on the fiber surface and video-recording length of fiber segments divided by the particles. Then, we simultaneously recorded length and stiffness changes, measured with sinusoidal fiber length changes (0.1% of fiber slack length L_0, 0.9 kHz), in tetanized single intact frog muscle fibers before, during, and after slow stretches, from sarcomere length 2.6 μm to sarcomere length 2.75 μm. The results are summarized in Fig. 115. The force in the fiber rose during stretch and after the end of stretch decayed to the same steady level, which was much higher than the level of isometric force at sarcomere length 2.75 μm. In contrast, muscle fiber stiffness showed an abrupt increase on application of stretch but started to decay while the stretch still continued to reach the same steady level, which was similar to the level attained during isometric force at sarcomere length 2.75 μm.

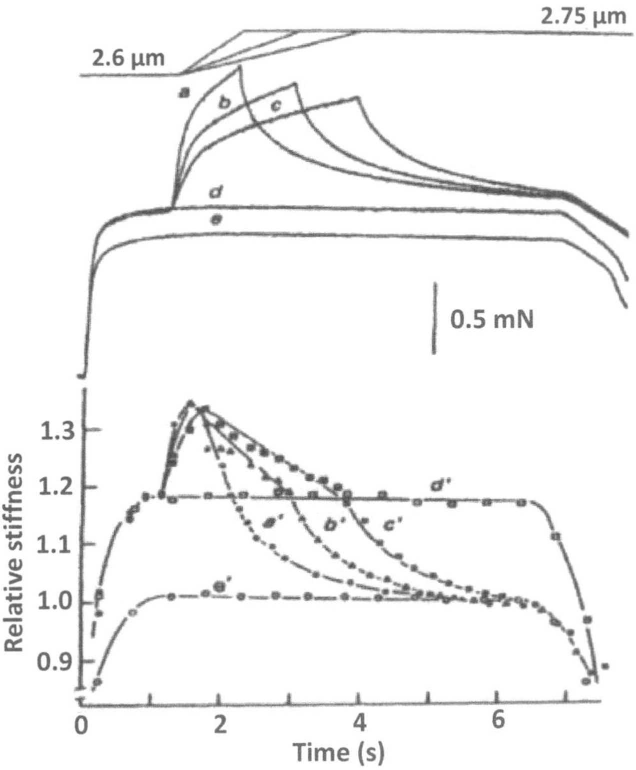

Figure 115 Force and stiffness changes in a single tetanized muscle fiber in response to slow stretches with three different velocities. (Upper records) Length and force changes in the fiber, which was first tetanized isometrically at sarcomere length 2.6 μm, and then stretched to sarcomere length 2.75 μm with three different velocities (traces a, b, and c). Traces d and e are force records of isometric tetanus at 2.6 and 2.75 μm sarcomere lengths, respectively. (Lower records) muscle fiber stiffness changes. Lines a′, b′, c′, d′, and e′ are stiffness changes corresponding to force records a, b, c, d, and e, respectively. From Sugi and Tsuchiya (1988).

The above unexpected finding can be explained as follows: (1) the initial abrupt rise of muscle fiber stiffness on application of stretch may result from the distortion of myosin heads when they are forcibly displaced in the direction opposite to that of muscle fiber shortening; (2) the subsequent decay of stiffness during stretch may reflect the decrease in the number of myosin heads interacting with actin filament with decreasing amount of overlap between actin and myosin filaments; and (3) the further decay of stiffness to the value equal to that during steady isometric force generation of the fiber at the stretched fiber length (2.75 μm sarcomere length) irrespective of velocity of stretch might result from the disappearance of distorted myosin heads at the beginning of stretch as they repeat attachment–detachment cycle with actin filaments.

If this explanation is more or less correct, the state of myosin heads reached after applied stretch is the same as that during isometric force generation at the stretched fiber length. Then a question arises why the force in the stretched fiber stays steady level much higher than that during isometric force generation at the stretched fiber length. One possible explanation is the storage of work done on the stretched muscle or muscle fibers during applied slow stretch. This possibility is discussed in the next section.

11.5 Time-Resolved X-Ray Diffraction Studies on Equatorial Reflections from Contracting Muscle during Slow Stretch

We explored possible answer to this puzzling question that muscle stiffness, which is believed to serve as a measure of the number of myosin heads involved in force generation, stays at a low level despite marked enhancement of isometric force attained after stretch, by using the technique of time-resolved X-ray diffraction. The results obtained are summarized in Fig. 116. The initial isometric force development was associated with the decrease in intensity of equatorial 1,0 reflection and the corresponding increase in intensity of 1,1 equatorial reflection, as reported by may authors (e.g., Huxley and Brown, 1967; Elliott et al., 1967; Haselgrove and Huxley, 1973), and both intensities reached a steady level when the isometric force reached a plateau. In response to the subsequent stretch, the force rose to a peak during stretch, and after completion of stretch started to decay to approach a steady level higher than the steady force before the application of stretch, I,e, stretch-induced enhancement of isometric force. On the other hand, the half-width of both 1,0 and 1,1 reflections did not change appreciably throughout the experiments.

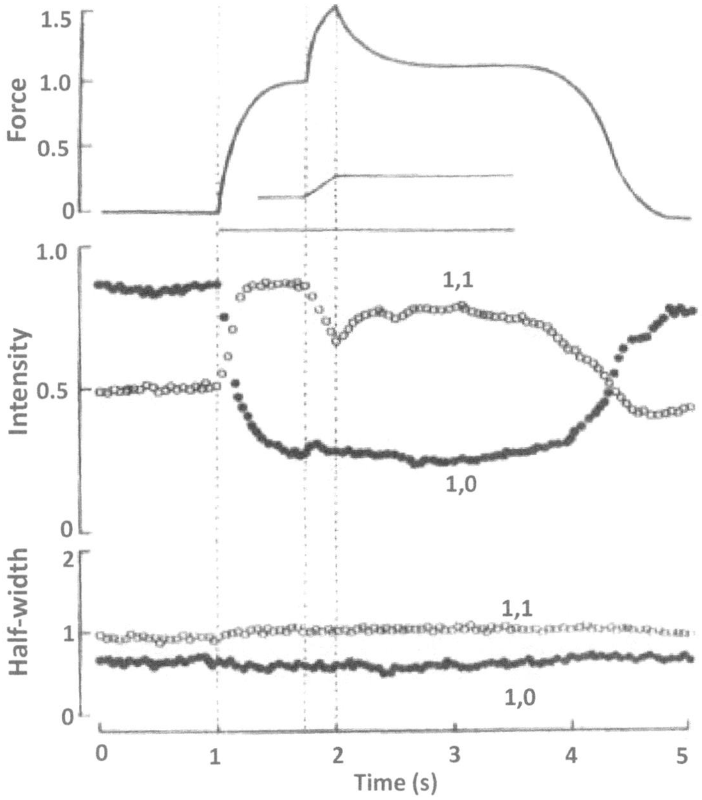

Figure 116 Effect of a slow stretch (4% of fiber slack length, L_0) on the force (upper trace), and on the intensity (middle traces) and half-width (lower traces) of the 1,0 (filled circles) and the 1,1 (open circles) equatorial reflections (middle and lower traces, respectively) from frog skeletal muscle. Force is expressed relative to the maximum isometric force, while intensity and half-width of equatorial reflections are expressed in arbitrary units. Each data point is an average of 20 tetani. Time course of stretch and period of tetanic stimulation are indicated under force trace. From Amemiya et al. (1988).

The equatorial X-ray diffraction pattern of skeletal muscle arises from the hexagonal myofilament lattice shown in Fig. 2 and has been studied intensively for many years. The 1,0 plane consists only of the myosin filament (Fig. 2, middle), while the 1,1 plane consists of myosin and actin filaments in a ratio of 1: 2 (Fig. 2, left and right). The decrease of the 1,0 reflection intensity associated with the increase of the 1,1 reflection intensity during isometric force development is interpreted as being due to the movement of the X-ray scattering mass, i.e., myosin heads, from the vicinity of myosin filaments to the vicinity of actin filaments (H. E. Huxley, 1968). Our experiments have shown that during stretch applied to the isometrically contracting muscle, the 1,1 intensity exhibited a marked decrease, which was followed by a partial recovery, whereas the 1,0 intensity did not change appreciably (Fig. 116). These results cannot be explained by change in the number of myosin heads attached to actin filaments.

A most plausible explanation is that a slow stretch applied to the isometrically contracting muscle causes the disordering of the 1,1 myofilament lattice without significant changes in the average interfilament spacing as shown diagrammatically in Fig. 117. Before stretch, the hexagonal myofilament lattice is well ordered, so that actin filaments are located in trigonal positions in the hexagonally packed myosin filaments (Fig. 117A). On application of stretch, actin filaments are displaced from their initial position without significantly changing their overall mean position in myosin filament lattice (Fig. 117B). The stretch-induced disordering of the myofilament lattice persists after the end of stretch, as indicated by the low 1,1 intensity level. The above idea is supported by the report of Matsuda and Podolsky (1986) that the myosin filament lattice is more stable than the actin filament lattice.

It seems possible that actin filaments are positioned within the myosin filament lattice by electrostatic repulsion forces between the two filaments. Since the electrostatic force is inversely proportional to the square of the interfilament distance, the stretch-induced actin filament displacement from the trigonal position in myosin filament lattice (Fig. 117B) may result in an increase in the overall repulsion forces within the whole myofilament lattice. It is well known that the myofilament lattice in the intact muscle or muscle fiber behaves as a constant-volume system. In the constant-volume myofilament lattice, the stretch-induced increase in the overall electrostatic repulsion forces between the filaments may produce an increase in the force required not only to get them closer together but also to maintain their disordered position.

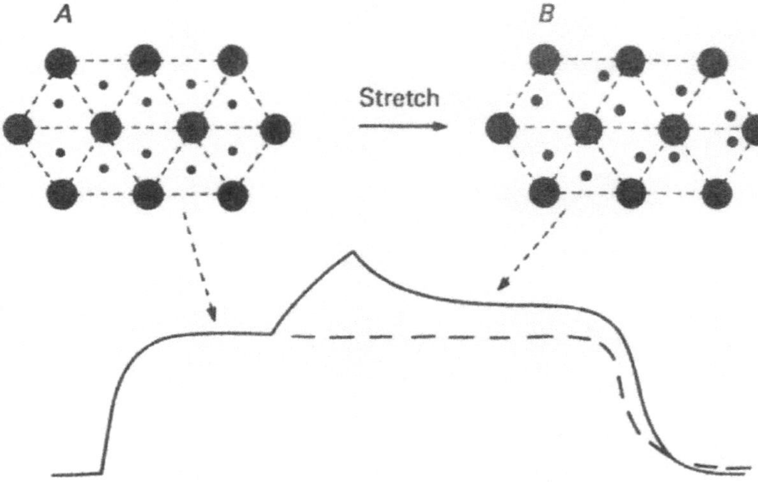

Figure 117 Possible disordering of myofilament lattice by stretch to cause force enhancement after stretch. From Amemiya et al. (1988).

Thus, the increased lateral electrostatic repulsion forces in the constant-volume myofilament lattice are converted into an increase in the isometric force, which shows up as stretch-induced enhancement of the isometric force. The explanation stated above might give an answer, at least in part, to the question in what form the work is stored in the 3D myofilament lattice.

Furthermore, if it is assumed that the increased overall electrostatic repulsion forces in the muscle transverse plane cannot be sensed as an increment of stiffness, measured by small fiber length changes, the stretch-induced enhancement of isometric force associated with decreased muscle fiber stiffness (Fig. 115) can be understood at least qualitatively.

11.6 Electron Microscopic Studies on Stretch-Induced Disorder of Myofilament Lattice

To confirm stretch-induced disorder of the myofilament lattice in stretched muscle, as suggested by time-resolved X-ray diffraction studies (Amemiya et al., 1988), intact single frog muscle fibers were fixed in OsO_4 at rest, during isometric tetanus, and after slowly stretching the tetanized fiber by ~13%. Figure 118 shows electron micrographs of the cross section of the fibers fixed in the three different states. It was difficult to see the degree of disordering among the three states by the observation of these micrographs by eye.

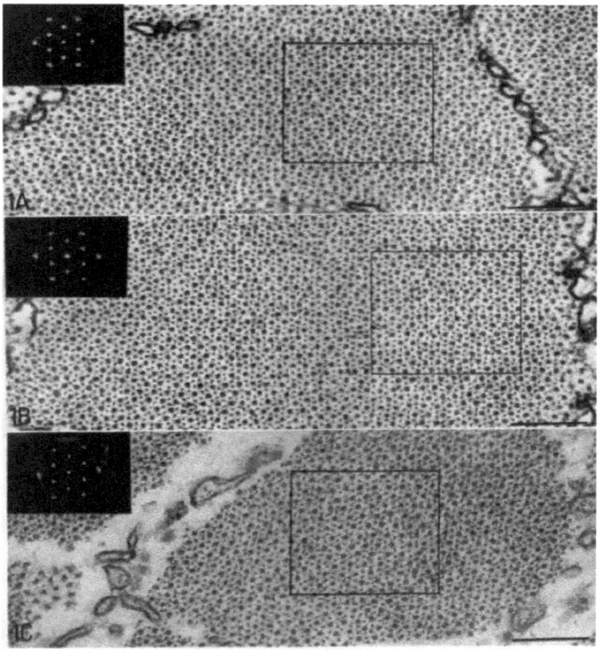

Figure 118 Electron micrographs showing cross sections of intact single frog muscle fibers fixed at rest (A), during isometric tetanus (B), and after a slow stretch (~13% of L_0) during tetanus (C). Fourier transform diffractograms obtained from square-shaped areas are also shown on the left of each micrograph. Bar, 0.2 μm (Suzuki et al., 1989).

To visualize the degree of disorder of the myofilament lattice with sixfold symmetry among the three different states, each electron micrograph shown in Fig. 118 was rotated in six steps (60° for each) around an arbitrarily chosen myosin filament. The superimposed electron microscopic images of myofilament lattice are shown in Fig. 119. The degree of disorder of the hexagonal myofilament lattice can be estimated by the distance from the center of rotation, over which the regular hexagonal myofilament lattice can be recognized. It was found that the myofilament lattice was most disordered in Fig. 119C, i.e., in the fibers fixed after stretch during tetanus, and most ordered in the fibers fixed during isometric tetanus without stretch. These results support the idea, already mentioned, that force enhancement by stretch results from stretch-induced disordering of myofilament lattice.

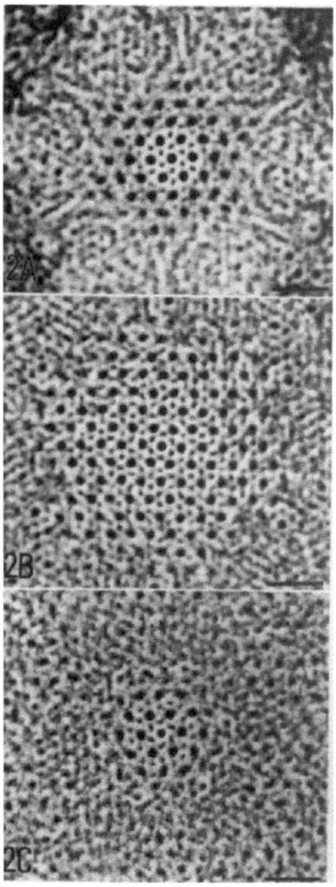

Figure 119 Superimposed images of the cross sections shown in Fig. 96, which are rotated around a myosin filament in 6 steps (60° for each). The images A, B, and C are obtained by rotating electron micrographs A, B, and C, respectively (Suzuki et al., 1988).

11.7 Measurement of Muscle Fiber Stiffness in Single Tetanized Muscle Fibers before, during, and after Slow Release

In contrast with slow stretches, during which force and stiffness in a tetanized muscle fiber showed different time courses (Fig. 115), muscle fiber stiffness was observed to change in parallel with the force in the fiber before, during, and after slow releases (Fig. 120). It has been reported that, during a slow release, the applied release was taken up by the two fiber end segments, while the length of central fiber segment remained unchanged (Julian and Morgan, 1979). We also examined length changes of the fiber segments during a slow release by putting carbon particles along the entire fiber length and recording length changes of fiber segments divided by the particles with a high-speed video system.

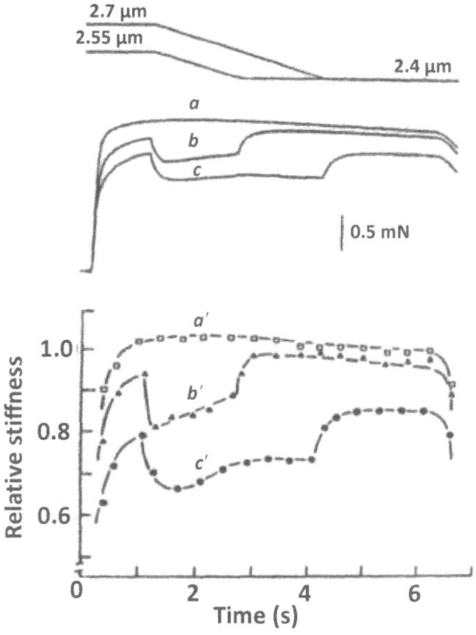

Figure 120 Force and stiffness changes in a single tetanized muscle fiber in response to slow releases. The fiber was first tetanized isometrically at 2.7 μm sarcomere length (traces b, b′) or at 2.55 μm sarcomere length (traces c, c′), and then slowly released to 2.4 μm sarcomere length. Traces a, a′ are obtained from isometric tetanus at sarcomere length 2.4 μm. From Sugi and Tsuchiya (1988a).

We confirmed that the applied fiber length change was mostly taken up in the two end segments, while the length of central segments did not change appreciably (Fig. 121) (Sugi and Tsuchiya, 1988). Since the force is uniform in every fiber cross section on a slow time scale, these results indicate that the deficit of force during and after a slow release produces the deficit of force not only in the two fiber end segments, which take up most applied release, but also in the middle fiber segments, the length of which does not change appreciably during release.

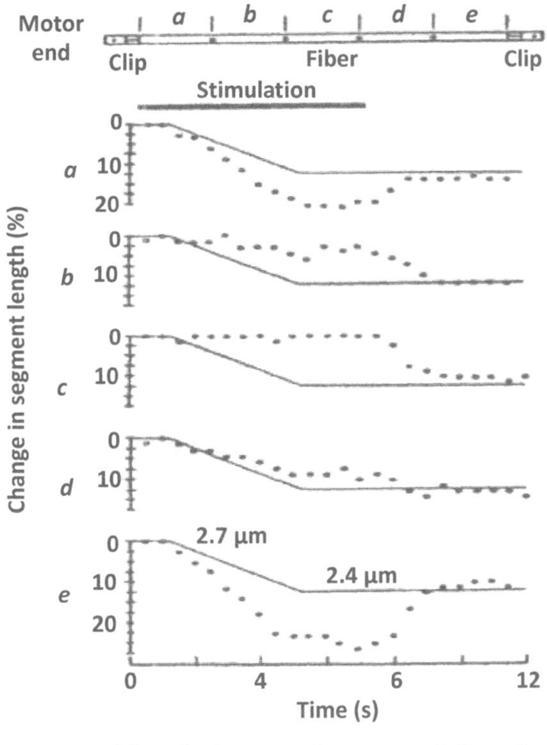

Figure 121 Segmental length changes of a tetanized fiber during a slow release from sarcomere length 2.7 μm to sarcomere length 2.4 μm. Note that the applied release is mostly taken up by two end segments (b and e) (Sugi and Tsuchiya, 1988).

We also performed time-resolved X-ray diffraction experiments on tetanized whole muscles during slow releases. As shown in Fig. 122, the intensity of both 1,0 and 1,1 equatorial reflections, as well as their half-width, did not change markedly during the applied slow release. At present, I have no definite idea to account for the mechanism underlying the force deficit after slow release.

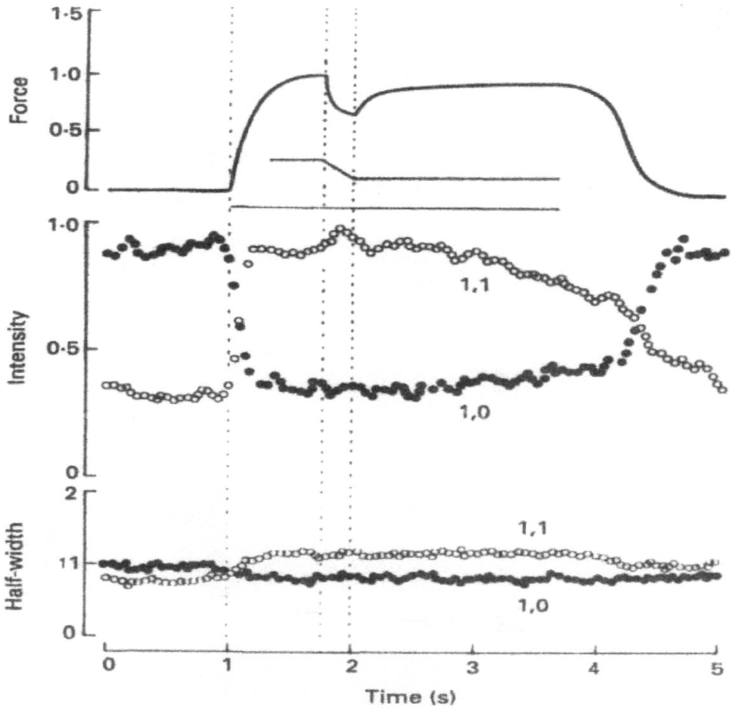

Figure 122 Changes in the intensity and half width of 1,0 and 1,1 equatorial reflections from a tetanized muscle during a slow release (4% of muscle length). Each data point is an average of 20 tetani (Sugi and Tsuchiya, 1988).

11.8 Longitudinal Stability of Sarcomere Length along the Fiber

In intact muscle fibers, sarcomere lengths near the tendinous ends are shorter and less extensible than those in the other regions (Huxley and Peachey, 1961), and therefore sarcomere lengths are not uniform when the fiber is made to contract isometrically with the two ends fixed in position. To obtain the relation between the sarcomere length and isometric force in tetanized muscle fibers, Gordon et al. (1966) used the length clamp method, in which the length of a central fiber segment, where the sarcomere length was uniform, was kept constant during isometric tetanus. Since the force is the same in every cross section of the fiber on a time scale not so fast to be comparable to that during transmission of mechanical impulse (Sugi and Kobayashi, 1983), it is possible to obtain the isometric force versus sarcomere length relation, and to correlate it to the amount of overlap between actin and myosin filaments within a sarcomere.

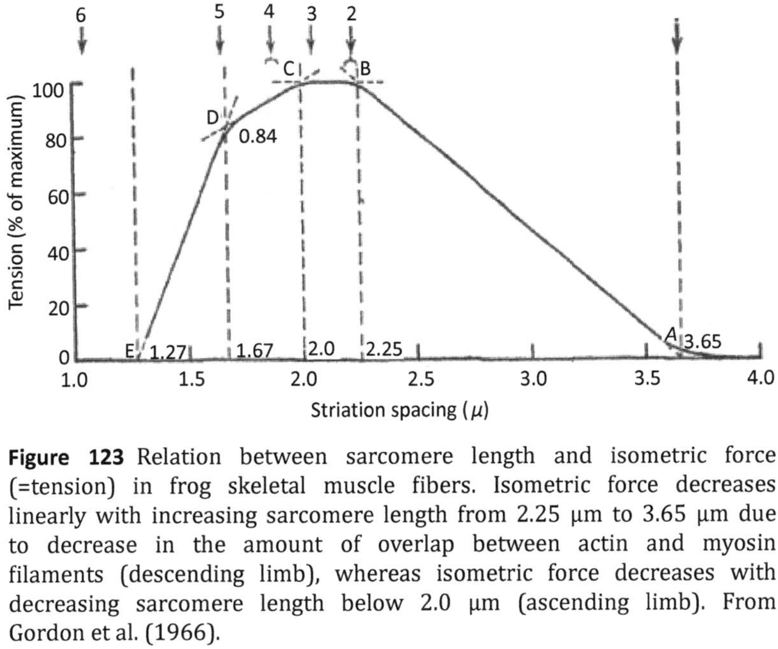

Figure 123 Relation between sarcomere length and isometric force (=tension) in frog skeletal muscle fibers. Isometric force decreases linearly with increasing sarcomere length from 2.25 μm to 3.65 μm due to decrease in the amount of overlap between actin and myosin filaments (descending limb), whereas isometric force decreases with decreasing sarcomere length below 2.0 μm (ascending limb). From Gordon et al. (1966).

As illustrated in Fig. 123, the sarcomere length versus isometric force relation consists of three parts; (1) the descending limb where the isometric force decreases linearly with increasing sarcomere length from 2.25 to 3.65 μm due to the decrease in the amount of overlap between actin and myosin filaments, i.e., decrease in the number of myosin heads that can interact with actin filaments, (2) the plateau where the isometric force is constant over sarcomere lengths from 2.0 to 2.25 μm (equal to the bare region width), and (3) the ascending limb where the isometric force decreases with decreasing sarcomere length below 2.0 μm due to double overlap of actin filaments and collision of the myosin filament against the Z-line (Gordon, Huxley and Julian, 1966).

The sarcomere length versus isometric relation is in itself unstable: many sarcomeres are connected in series within a muscle fiber. If some sarcomeres generate less force than that generated by other sarcomeres during isometric force generation in the fiber, they will be stretched by other sarcomeres, and generate less and less force along the descending limb to eventually reach the length at which it generates no force. This will result in a considerable non-uniformity in sarcomere length along the fiber, and eventually the fiber becomes unable to generate force externally. Conversely, if some sarcomeres generate more force than others, they will shorten by stretching other sarcomeres and will generate less and less force along the ascending limb. In both cases, the fiber cannot generate steady isometric force, unless the sarcomere length versus isometric force relation is modified.

Both the force enhancement by stretch and the force deficit by release are useful to keep the longitudinal stability of sarcomeres along the entire fiber length, although both mechanisms remain to be investigated.

Chapter 12

Reminiscences of My Early Work: I. Summation of Force Responses in Muscle

The subsequent chapters are added to this volume by request of the publisher and deal with my early work on muscle. At first, I felt somewhat reluctant to describe the experimental work done in my young days, since the topics to be described in this chapter have already been clarified at least qualitatively up to the molecular level. When I started writing this chapter, however, I was gradually interested in describing my early work. At that time, almost everything about muscle contraction remained mysterious, so that I could design experiments freely with a hand-made experimental setup, which may seem to be simple and too primitive for young physiologists, who are accustomed to using expensive sophisticated experimental apparatus. Regardless of whether young people respond positively or negatively to this chapter, I was excited with what I achieved with my simple, mostly hand-made equipment. Frankly speaking, in my life as an investigator, I was most happy at that time compared with my subsequent period of research work with many coworkers and expensive apparatuses.

Mysteries in Muscle Contraction: Evidence against Current Dogmas
Haruo Sugi

ISBN 978-981-4774-48-2 (Hardcover), 978-1-315-10191-0 (eBook)
www.panstanford.com

I started my research work as a muscle physiologist in 1962 in the Department of Physiology, University of Tokyo. At that time, it was a complete mystery how action potential could elicit muscle contraction, and many physiologists had an ambiguous idea of "excitation," based on the all-or-none action potential and all-or-none twitch in muscle, and they thought that both action potential and muscle contraction originated from the common underlying mechanism; in other words, action potential was electrical expression of excitation, while its mechanical expression was contraction. Based on this idea, many investigators made non-productive work over many years.

12.1 A. V. Hill's Concept of Active State: A Wrong Dogma

In 1949, Hill reported that if a muscle was stretched rapidly by ~15% immediately after application of a current pulse to elicit an isometric twitch, the force in muscle rose to a level equal to the maximum tetanus tension, P_0 (Fig. 124). Hill interpreted this result as evidence for very rapid transition from rest to full activity of the contractile component immediately after application of a current pulse. On the basis of his two-component model, in which muscle consists of the contractile component (CC) and the SEC connected in series (Fig. 2B), Hill reached a conclusion, which can be summarized as follows: (1) Immediately after application of a current pulse, the CC in muscle is put into fully activated state, i.e., active state; (2) the active state is all-or-none in nature; (3) the rapid, all-or-none development of the active state cannot be seen externally, as the CC should stretch the SEC internally with velocities determined by the force–velocity relation; (4) the duration of the active state is short and starts to decay before the externally measured force is still rising; (5) as the result, the active state decays considerably when the isometric twitch force reaches a peak, and at this time point, the magnitude of the active state is equal to the peak isometric twitch force.

In the light of the current knowledge that muscle twitch is associated with a single action potential, propagating along the fiber surface, Hill's concept of active state was wrong in postulating all-or-none nature of the contractile mechanism, which was later proved to be associated with chemical reactions of contractile proteins. Nevertheless, the active state concept was regarded as the central dogma in the research field of muscle mechanics over many years.

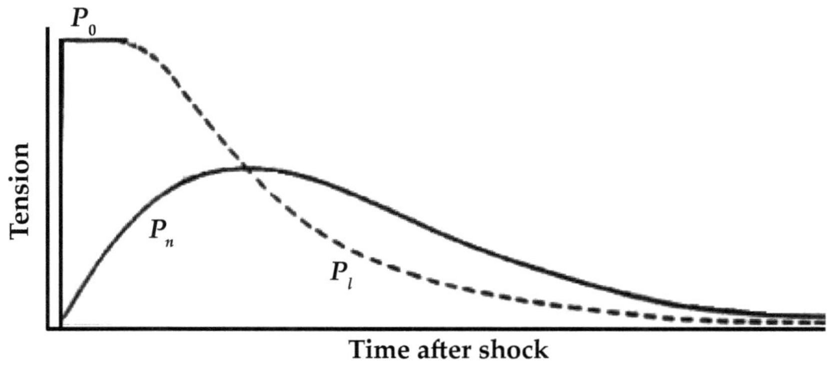

Figure 124 Diagram of experiments, from which Hill obtained the concept of active state in stimulated muscle. Trace P_n indicates time course of an isometric twitch, while trace P_i indicates time course of active state in the contractile component (CC), obtained by rapidly stretching a muscle immediately after application of current pulse by 15%. Note that on stretch, the force in muscle rises to a peak equal to that of the maximum isometric tetanus, P_0 (Hill, 1949).

12.2 Muscle Contraction Is Produced by Depolarization of Surface Membrane

In skeletal muscle, contraction is normally initiated by action potential propagating along the fiber surface, and this led Hill to think of his active state concept. If action potential is blocked, localized contraction, i.e., contracture, still takes place by various means, indicating that the contractile mechanism is independent of action potential taking place at the surface membrane.

By using single muscle fibers, in which action potential was blocked by replacing external Na^+ with choline, Hodgkin and Horowicz (1960) showed the relation between external K^+ concentration, i.e., membrane potential, and force development (Fig. 125). The fiber developed force when the surface membrane was depolarized by ~40 mV, and with further depolarization force increased steeply to a peak. The critical membrane depolarization for the development of force was nearly the same as that for the development of action potential. Further studies by several investigators have shown that it is membrane depolarization *per se* that triggers muscle contraction, but not the flow of current in muscle fibers.

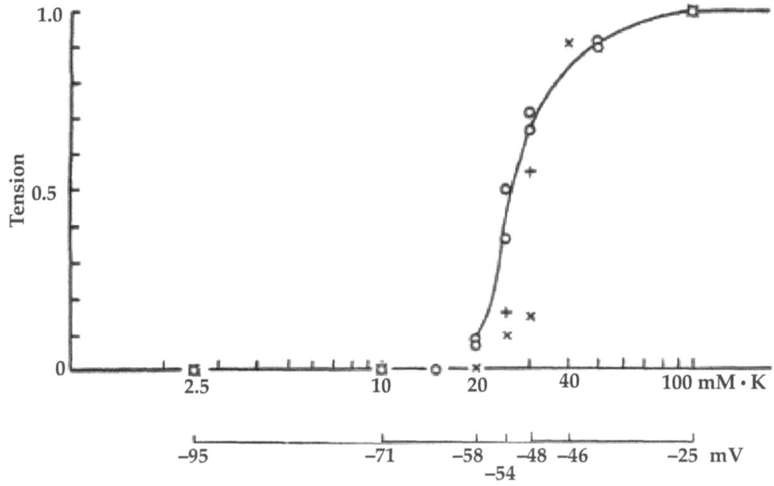

Figure 125 Relation between force development and external K^+ concentration (or membrane potential) of a single muscle fiber. From Hodgkin and Horowicz, 1960).

12.3 Transverse Tubular System Transmitting Surface Membrane Depolarization into the Fiber

In 1949, Hill also pointed out that the rapid onset of active state, as evidenced by the rapid increase in resistance of muscle against applied stretch (Fig. 124), could not be explained as being due to diffusion of molecules or ions liberated at the surface membrane throughout the interior of muscle fibers, suggesting the presence of some organized structures that rapidly transmit the influence of action potential deep into the fiber.

To answer this question, A. F. Huxley and Taylor (1958) performed experiments in which they passed current through a micropipette (diameter, ~μm) in contact with frog muscle fiber surface, so that a limited area of the surface membrane was locally depolarized. They found that when the I-band area including the Z-line was depolarized, contraction localized under the micropipette was observed, while no contraction was produced when only the A-band area was depolarized. The local contraction was not of all-or-none type but increased with increasing depolarization. These results clearly indicated the presence of internal structures, localized in the I-band region, and transmit the effect of surface membrane depolarization into the interior of the fiber.

After the work of A. F. Huxley and Taylor (1958), it became clear that the surface membrane of frog muscle fibers has many invaginations, which are continuous with the membrane of the transverse tubular system (T-system), forming a tubular network around each myofibril. The location of the T-system differs from animal to animal; in frog muscle fibers, the T-system is located at the level of Z-band, while in mammalian muscle fibers, the T-system is located at the level of boundary between A- and I-bands.

The above progress in the research field, starting from the concept of active state to the finding of the T-system, was just in progress when I started working on muscle in 1962.

12.4 Summation of Isometric Twitch: Argument about the Baseline from Which the Second Twitch Starts

When I was appointed an instructor in physiology, I read many published papers on muscle contraction and was interested in the mode of summation of two successive isometric twitches as shown in Fig. 101. For estimating the amplitude of the second twitch, which starts after the first one, Hartree and Hill (1921) subtracted the first twitch from the fused mechanical response, on the assumption that the twitch fusion was algebraical in character.

I felt reluctance to their algebraical subtraction method. According to Hill's active state concept, externally recorded isometric force results from the internal shortening of the CC by stretching the SEC. If this concept is correct, the second twitch must produce additional stretch of the SEC; in other words, the second twitch should start from the force level attained by the first twitch. As a matter of fact, Fulton (1925) thought that the second twitch started *de novo* from a new baseline, at which the first twitch force was at the time. The two different methods of estimating the amplitude of the second twitch are shown in Fig. 126E. Later, I became aware that Hill changed his mind and stated in one of his papers published after 1949, that his past estimation of the second twitch by algebraical subtraction was invalid.

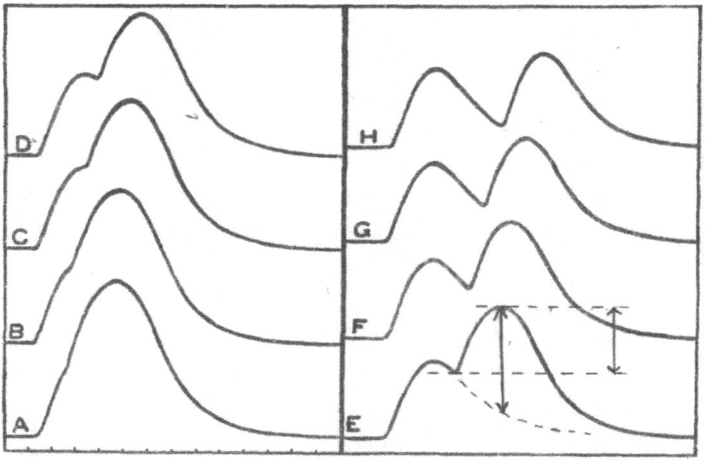

Figure 126 Summation of isometric twitch by two successive stimuli applied to a muscle. Intervals between two stimuli are 24, 32, 40, 48, 57, 69, 77 ms in A, B, C, D, E, F. G. H, respectively. In E, two different methods of estimating the second twitch are illustrated by arrows and broken lines. From Ruch and Fulton (1960).

12.5 Summation of Force Response in Crayfish Muscle Fibers

12.5.1 Observations on Vertically Suspended Muscle Fibers

As can be seen in Fig. 126, the interval between two successive stimuli cannot be made shorter than the absolute refractory period of action potential. At that time, an experimental idea that occurred to my mind was to use crayfish muscle fibers, which do not generate all-or-none action potential and exhibit graded contractions depending on the frequency of junction potentials arising from many neuromuscular junctions distributed along the fiber length.

To stimulate crayfish muscle fibers uniformly along their entire length without propagated action potential, it was necessary to put them in a transverse electric field, in which stimulating current flows into the fiber across the fiber membrane facing the anode, thus producing membrane hyperpolarization, and then flows out of the fiber across the fiber membrane facing the cathode, thus producing membrane depolarization.

Before starting experiments to examine the summation of isometric contractions, I felt it necessary to confirm by eye whether contraction actually occurs only one side of the fiber where surface membrane was depolarized or not. For this purpose, a single crayfish muscle fiber (diameter, 300–800 μm, slack length 0.5–1 cm) was isolated from the claw adductor of crayfish, with a piece of tendon attached to one end and a piece of shell left at the other end. A single crayfish muscle fiber was suspended vertically in an experimental chamber with Ag–AgCl electrodes covering the two opposite walls to apply transverse current pulse along the entire fiber length (Fig. 127A). The strength of the transverse electric field was ~3 V/cm (Sugi and Kosaka, 1964).

In accordance with expectation that contraction is localized in one side of the fiber where the stimulating current flows out of the fiber to produce membrane depolarization, the suspended fiber bent quickly in the form of arc in response to a brief transverse current pulse (duration, 1–5 ms), in such a way that the cathodal side of the fiber lay inside the arc (Fig. 128B). Unexpectedly, however, if the current pulse was prolonged to more than several tens of milliseconds, the degree of fiber bending decreased gradually, and eventually the fiber shortened uniformly (Fig. 128C). This unexpected finding indicated that contraction, first produced at the cathodal side of the fiber, may gradually spread toward the anodal side where the membrane is hyperpolarized. When the fiber was deteriorated after repeated application of current pulse, the fiber no longer contracted uniformly (Fig. 128D).

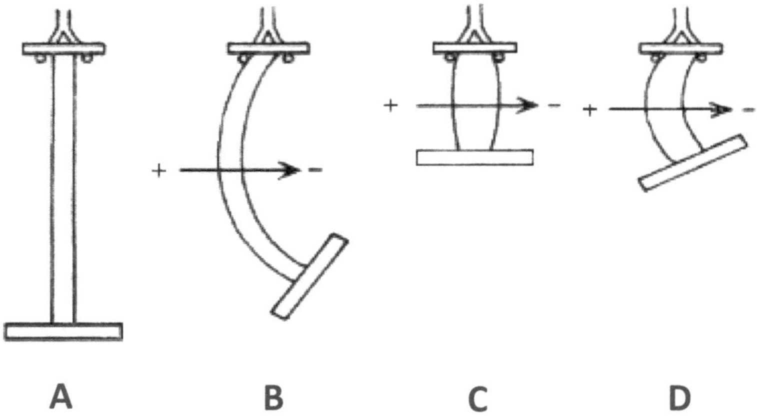

Figure 127 Diagrams showing the response of a suspended crayfish muscle fiber to transverse current pulses. Arrows indicate current direction. (A) Fiber at rest. (B) Fiber bending in response to a brief current pulse. (C) uniform fiber shortening in response to a prolonged current pulse. (D) Non-uniform shortening of a deteriorated fiber in response to a prolonged current pulse. From Sugi and Kosaka (1964).

12.5.2 Summation of Force Response by Two Successive Transversely Applied Current Pulses

12.5.2.1 Two successive current pulses of opposite directions

I was excited with the finding obtained from suspended muscle fibers and performed experiments in which the fiber was held horizontally by fixing one end in position, while the other end connected to a force transducer, in an experimental chamber to be subjected to transversely applied current pulses (Fig. 128).

First, I applied two successive brief current pulses of opposite directions to produce contraction at both sides of the fiber in succession and recorded the summation of the two brief isometric responses. As the two sides of the fiber are mechanically connected in parallel with each other, the two responses were expected to sum algebraically. Figure 129A presents the superimposed records showing the summation of two successive isometric responses, produced by two successive current pulses of opposite direction. In accordance with the above expectation, the magnitude of the fused force response increased with decreasing current pulse interval, and the analysis of force records actually showed the algebraical nature of summation of the two successive responses (Fig. 130).

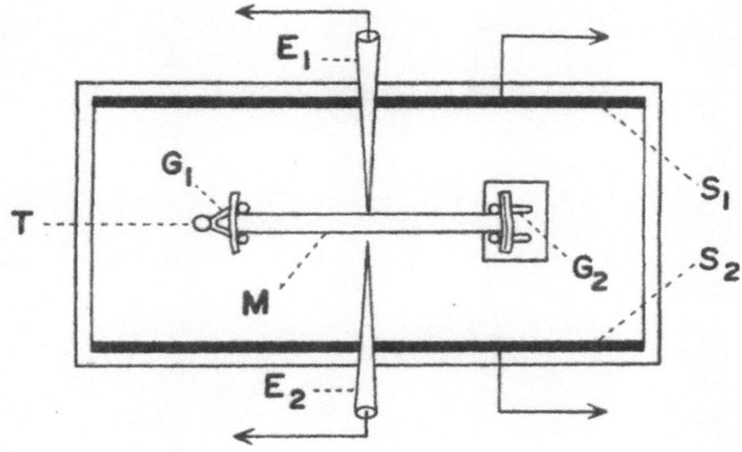

Figure 128 Diagram of experimental arrangement. A single crayfish muscle fiber was held horizontally by fixing one end (G_2) in position while the other end (G_1) connected to a force transducer. Two opposite walls of the chamber was covered by Ag-AgCl plate electrodes (S_1 and S_2) to apply transverse DC field of ~3 V/cm. Membrane potential changes were recorded with two microelectrodes (E_1 and E_2). From Sugi and Kosaka (1964).

As can be seen in record B in Fig. 129, the force response rose to a peak, which was well maintained as long as the current pulse went on; if the direction of the prolonged pulse was reversed, its magnitude did not increase at all but showed a slight transient decrease, which was followed by restoration of the initial force level. The maintenance of the steady force level during prolonged current pulse could be repeated several times. After this period, the fiber started to deteriorate, so that the force level decreased with time, and the reversal of current direction produced additional force (Fig. 129C).

I was again excited with the results shown in Fig. 129B, since it seemed to correspond to the observation of suspended fibers that during the application of prolonged current pulse, the fiber shortened uniformly (Fig. 127C). In both cases, the contraction first initiated by one side of the fiber, where the membrane is depolarized, spreads toward the other side, where the membrane is hyperpolarized.

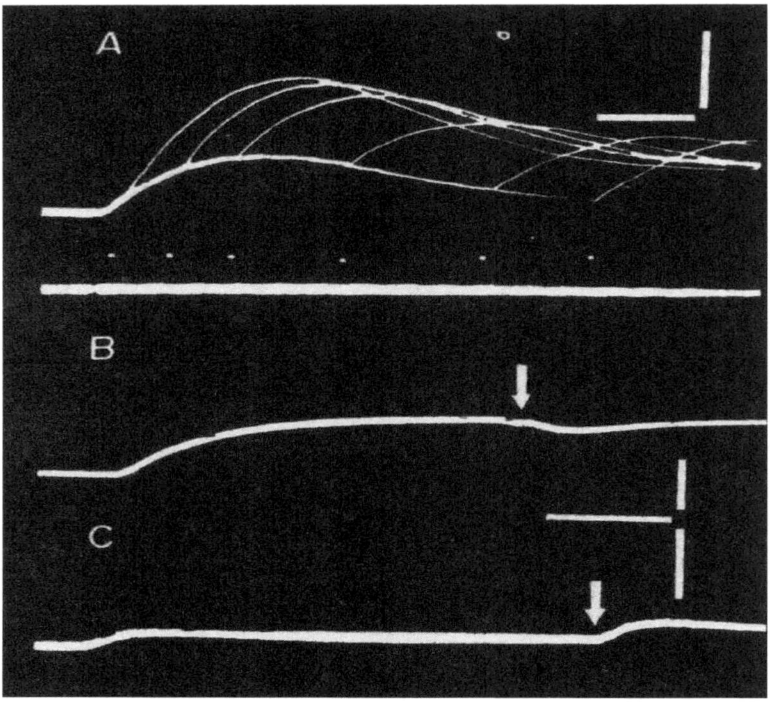

Figure 129 Isometric responses of a single crayfish muscle fiber to transversely applied current pulses. (A) Superimposed force records showing summation of two brief isometric responses produced by two successive brief current pulses (2 ms, 3 V/cm) given at various intervals. Note that the magnitude of fused isometric response increases with decreasing current pulse intervals. (B and C) Force record in response to a prolonged current pulse (3 V/cm), which was applied as long as force response continued. Record B was obtained from a fresh muscle fiber, while record C was obtained from a deteriorated muscle fiber. Note that, in record B, the magnitude of force response stays constant during the application of prolonged current pulse, and when the direction of current pulse was reversed (downward arrow), the force falls slightly and then restores the same steady magnitude. Force calibration, 500 mg in A, 3g in B, and 1g in C. Time calibration, 50 ms in A and 0.5 s in B and C. From Sugi and Kosaka (1964).

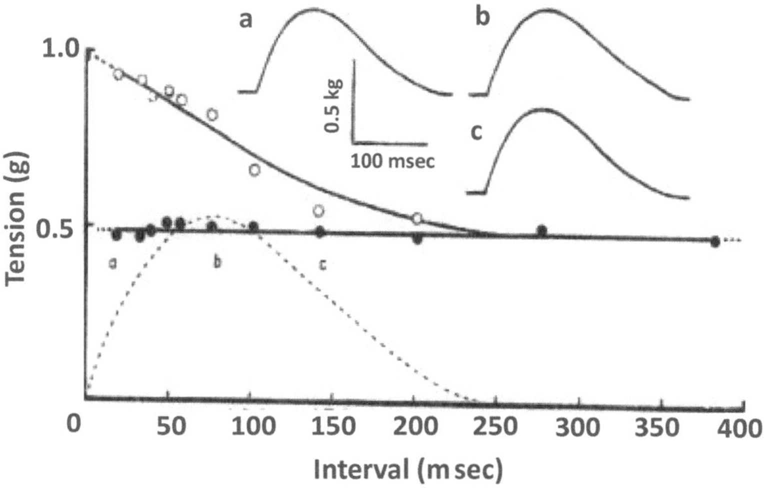

Figure 130 Summation of force responses produced by two successive brief current pulses (2 ms, 3 V/cm). Open circles indicate the magnitude of summated response, while filled circles indicate the magnitude of the second force response, obtained by algebraically subtracting the first force response from the fused one. Note that the magnitude of the second response is constant irrespective of the current pulse interval. Curves a, b and c are time course of the second response obtained by subtracting the first response (dotted line) from the fused one (Sugi and Kosaka, 1964).

12.5.2.2 Two successive current pulses of the same direction

In Fig. 131, superimposed force records show the mode of the summation of force responses produced by two successive current pulses of the same direction. In records A, the two pulses are the same in duration (2 ms, 3 V/cm), while in records B and C, the two pulses are different in duration (1 and 6 ms) in the first and the second pulse in B, while the sequence is reversed in C. Interesting features of summation are as follows: (1) The summated response is maximum when the second response starts at the rising phase of the first force response, (2) the magnitude of summated response is larger when long pulse comes after the short pulse, and (3) in all cases, the increment of force by the second response decreases abruptly as the pulse interval is made very short.

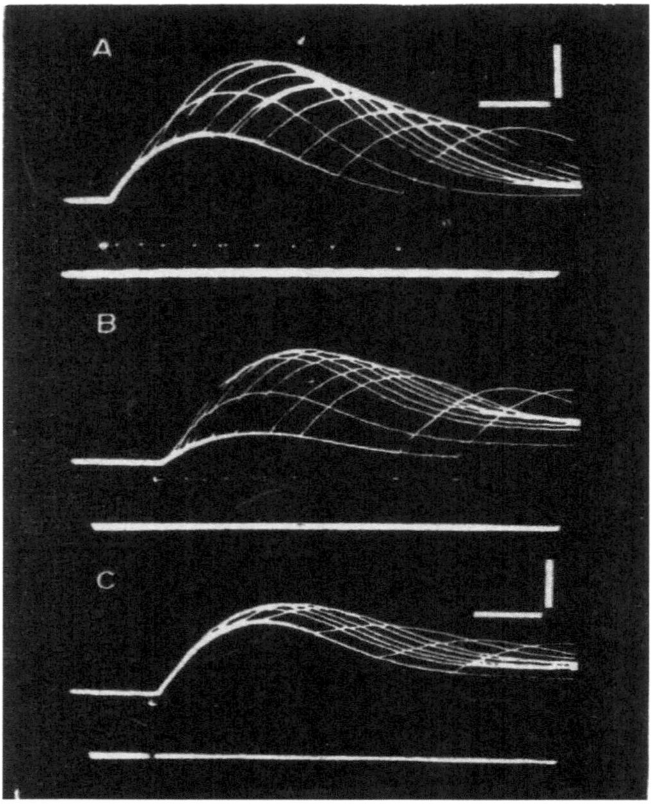

Figure 131 Superimposed records showing summation of force responses by two successive current pulses (3 V/cm) of the same direction. In A, the first and the second pulse had the same duration (2 ms). In B, the duration of the first and the second pulse were 1 and 6 ms, respectively. In C the sequence is reversed. Note that the maximum magnitude of summated response is much larger when shorter current comes first (B) than when longer current pulse first (C). Calibration, 500 mg and 50 ms (Sugi and Kosaka, 1964; Sugi and Kosaka, 1964).

Figure 132 shows the analysis of the summation of force responses by two successive pulse of the same duration and direction (2 ms, 3 V/cm). At that time, the results shown in Fig. 132 appeared too complicated to me, and I made no further experiments on the summation of force responses. Instead, I started to perform another type of experiments to observe the spread of contraction in muscle fibers in the transverse direction under a light microscope with unexpected findings.

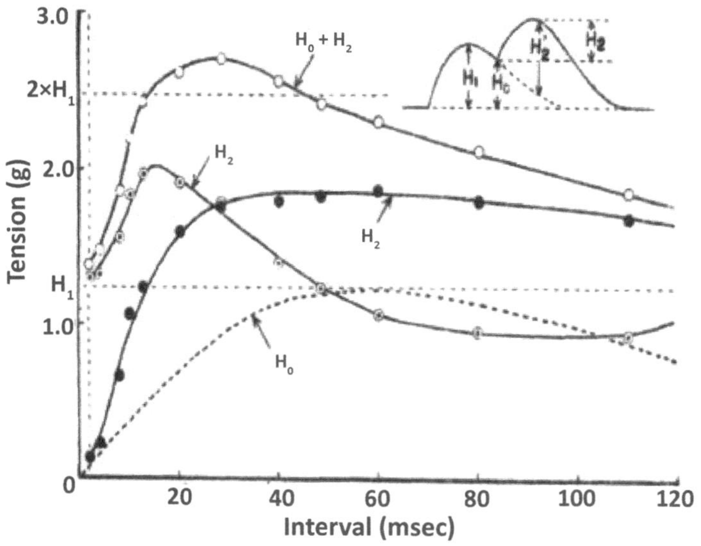

Figure 132 Summation of two force responses produced by two successive brief pulses (2 ms, 3 V/cm) of the same direction. Dotted line shows time course of the first response. Filled circles show magnitude of the second response measured by algebraical subtraction of the first response from the summated one, while open and dotted circles indicate magnitude of summated response and that of the second response measured from the baseline equal to the force, from which the second response starts. From Sugi and Kosaka (1964).

Chapter 13

Reminiscences of My Early Work: II. Transverse Spread of Contraction Initiated by Local Membrane Depolarization in Single Muscle Fibers

13.1 Crayfish Muscle Fibers

Since I performed experiments with Dr. Kosaka on the summation of force responses, using single crayfish muscle fibers, it was natural to start local activation experiments with Dr. Ochi using the same material, though we later used single frog muscle fibers.

13.1.1 General Features of Local Membrane Depolarization

Single muscle fibers (major diameter, 70–200 μm) isolated from superficial abdominal extensor muscle of the crayfish were mounted horizontally with both ends fixed in position. The surface membrane was locally depolarized by placing the tip of a glass pipette (diameter, 20–120 μm, filled with experimental solution) in contact with the fiber surface and applying current pulses to the pipette to make its interior go negative with respect to the

Mysteries in Muscle Contraction: Evidence against Current Dogmas
Haruo Sugi

ISBN 978-981-4774-48-2 (Hardcover), 978-1-315-10191-0 (eBook)
www.panstanford.com

surrounding medium. The contact resistance of the pipette, i.e., the increase of pipette resistance when its tip was brought into contact with the fiber surface, was measured by a Wheatstone bridge. The value of contact resistance ranged from 20 to 150 kΩ. The magnitude of membrane potential change across the surface membrane under the pipette was calculated by multiplying the contact resistance by the applied current.

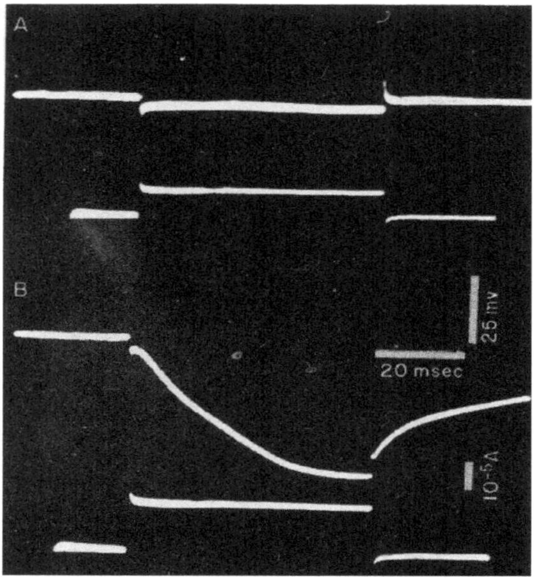

Figure 133 Change in membrane potential (upper records) of a crayfish muscle fiber during application of current (lower records) to the pipette in contact with the fiber surface. (A) Records obtained during application of a current pulse producing membrane depolarization of 150 mV. Note no change in membrane potential is produced except for a small artefactual deflection. (B) Records obtained during application of a current pulse producing membrane depolarization of 300 mV. Note hyperpolarization is produced as a result of reduced membrane resistance under the pipette (Sugi and Ochi, 1967a).

To ascertain whether the change in membrane potential produced by current pulses to the pipette was actually localized under the pipette, an intracellular glass capillary microelectrode was inserted into the fiber at a point close to the pipette. As shown in Fig. 133A, little or no membrane potential changes were recorded during the application of current pulse producing local membrane depolarization up to 200–300 mV, indicating that changes in membrane potential were actually localized under the pipette without appreciably changing membrane potential elsewhere.

Meanwhile, strong current pulses producing local membrane depolarization above 200–300 mV caused marked irreversible reduction of membrane resistance (to less than 100th) under the pipette, as evidenced by membrane hyperpolarization recorded at the fiber surface membrane around the pipette (Fig. 133B). The resulting considerable current flowing into the interior of the fiber around the pipette, however, still produced localized current flowing out from the interior of the fiber under the pipette.

Photomicrographs of contraction were recorded at any chosen moment after onset of a depolarizing current pulse with a 35 mm camera and a xenon lamp flash device (flash duration, <1 ms). Contractions were also recorded with a 16 mm cine-camera at 35–50 frames/s.

13.1.2 Inward Spread of Contraction Initiated by Moderate Membrane Depolarization

When the magnitude of local membrane depolarization exceeded 20–50 mV, a reversible local contraction was produced under the pipette (Fig. 134, see also Figs. 135 and 136). The contraction, i.e., the shortening of sarcomere length, occurred first at the upper edge of the fiber facing the pipette and spread inward. The extent of inward spread of contraction increased with increasing magnitude of local membrane depolarization.

The extent of inward spread of contraction for a given amount of depolarization also increased with increasing diameter of pipette for local activation. With a large pipette with a diameter nearly half as large as the fiber diameter, the resulting local contraction spread not only inward but also transversely across the center of the fiber (Fig. 135), though it was not possible to make the contraction spread all the way across the fiber with moderate membrane depolarization not to cause irreversible reduction of membrane resistance.

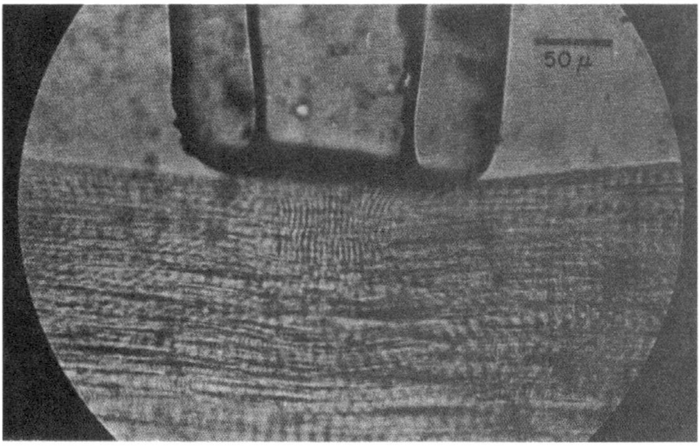

Figure 134 Local contraction initiated by membrane depolarization of 200 mV by a current pulse of 500 ms duration (Sugi and Ochi, 1967a).

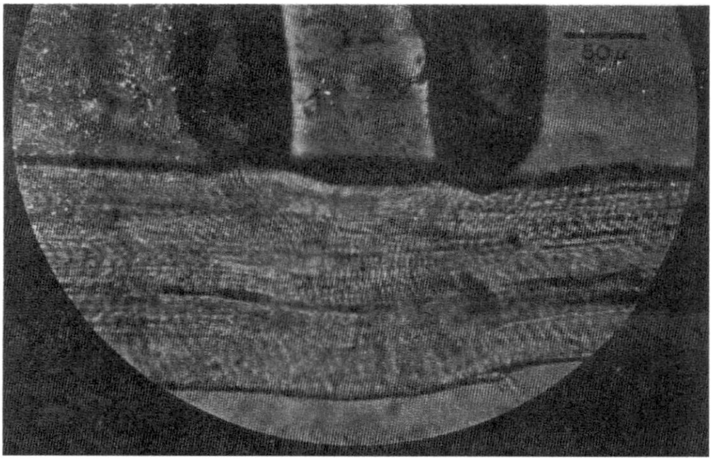

Figure 135 Local contraction spread across the center of the fiber, initiated by membrane depolarization of 180 mV (duration, 1 s) (Sugi and Ochi, 1967a).

13.1.3 Summation of Local Contraction Initiated by Moderate Membrane Depolarization

Figure 136 shows photomicrographs of local contractions initiated by moderate membrane depolarization (150 mV) of 200 ms duration (B), 50 ms duration (C), and two successive 50 ms duration given at an interval of 100 ms (D), together with the photo-micrograph of resting fiber (A). It can be seen that the local contraction produced by two successive 50 ms depolarizations (D) is just as large as that produced by a 200 ms depolarization (B) with respect to both the extent of sarcomere shortening and that of the inward spread of contraction, while the local contraction produced by a 50 ms depolarization (C) is much smaller than the other two types of local depolarization.

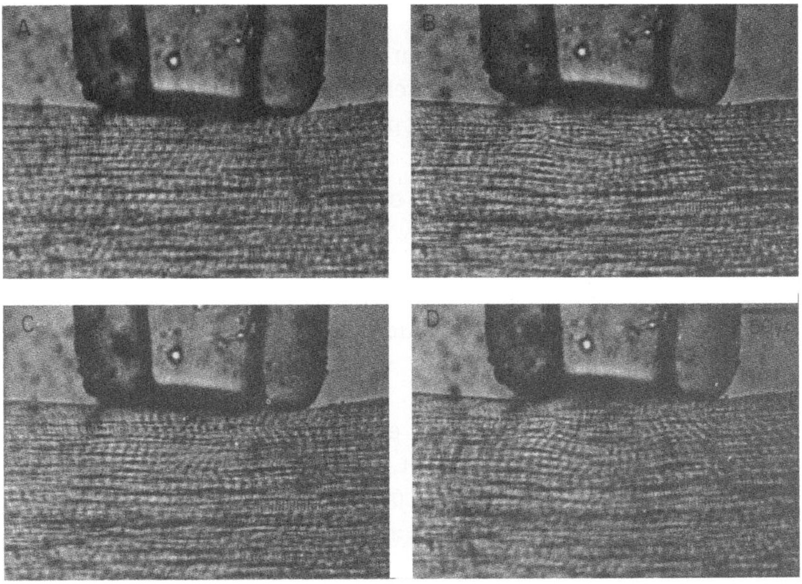

Figure 136 Effect of two successive brief depolarizations as compared with a long depolarization. (A) Resting fiber. (B) Local contraction by a 200 ms depolarization. (C) Local contraction by a 50 ms depolarization. (D) Local contraction by two successive 50 ms depolarizations (Sugi and Ochi, 1967a).

The result that two successive 50 ms depolarization given at an interval of 100 ms produce the same extent of local contraction as does a single 200 ms depolarization indicates that intermittent brief depolarization is more effective in producing local contraction than a single prolonged depolarization. These phenomena indicate that the effect of depolarization in producing contraction decreases with time during the application of a prolonged depolarization, probably due to the "inactivation" of the process liking surface membrane depolarization and Ca^{2+} release from the sarcoplasmic reticulum.

This idea is supported by our previous report on the summation of force response by two successive stimuli in single crayfish muscle fibers (Sugi and Kosaka, 1964); in Fig. 131, it can be seen that the summation of force response by two successive current pulses produces much larger summated response if the short current pulse comes ahead of the long one (B) than if the sequence is reversed (C); namely, the short current pulse can produce much less force if it is applied after long current pulses, than if it is applied without preceding long current pulse. This phenomenon may be related to the accommodation of excitation–contraction coupling process to be described later.

13.1.4 Transverse Spread of Local Contraction Initiated by Strong Current

In response to a strong current pulse producing depolarization >200–300 mV, crayfish muscle fibers exhibited another type of local contraction definitely different from that to moderate depolarization.

Figure 137 is a typical example showing the nature of reversibility of the response. It can be seen that the contraction spreads transversely from the upper edge of the fiber in contact with the pipette to the lower edge without spreading along the fiber axis. This type of local contraction spreading across the fiber could be initiated even with brief current pulses <1 ms, provided its strength was sufficient, though it was necessary to use pipettes with diameters >50 μm.

Closer microscopic observation of this type of local contraction revealed that contraction was localized around the whole fiber perimeter, but not at the center of the fiber. As shown in Fig. 138, the sarcomere shortening is observed at both the upper and lower edges, but not at the fiber middle region. This indicates that the contraction spreads around the fiber perimeter but not across the center of the fiber.

As already mentioned, the application of a strong current pulse markedly reduced the membrane resistance under the pipette, thus providing a pathway for large current flowing into the fiber. Once the membrane resistance was reduced, the amount of current to initiate contraction spreading around the fiber perimeter was very much reduced, indicating that it was the inward current flowing around the pipette that initiated this type of contraction. Based on the current knowledge that muscle contraction is initiated by Ca^{2+} released from the sarcoplasmic reticulum (SR) (Ebashi and Endo, 1968), and that the influence of surface membrane depolarization is transmitted along the transverse tubules (T tubules), which is continuous with the surface membrane (Huxley and Taylor, 1958), the reduced membrane resistance under the pipette provides a route for strong current, which flows into the fiber across the surface membrane around the pipette, and flows out at the fiber surface under the pipette. As a result, strong current may flow across the T tubule membrane, which are continuous with the surface membrane under the pipette, thus producing a large depolarization of the T tubule membrane, to initiate a regenerative process causing local contraction spreading around the fiber perimeter.

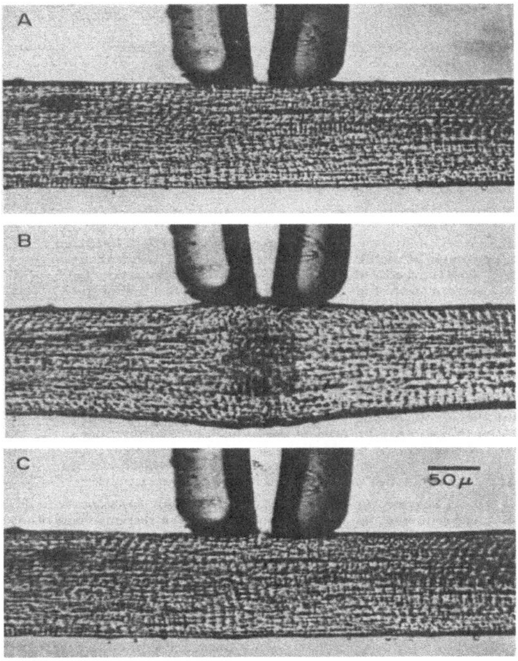

Figure 137 Selected frames from a cinefilm showing the fiber before (A), during (B), and after (C) a local contraction initiated by a strong current pulse (duration, 5 ms) producing depolarization of 450 mV. Note that the contraction spreads from the upper to the lower edge of the fiber (Sugi and Ochi, 1967a).

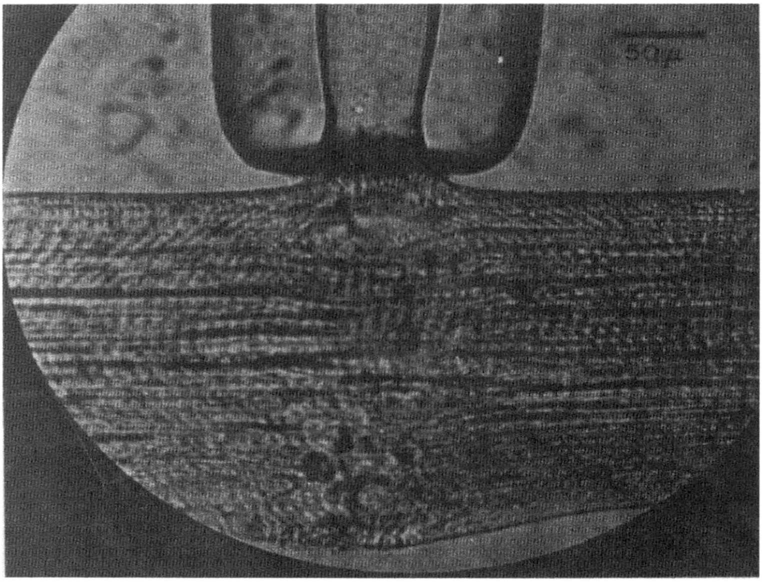

Figure 138 Photomicrograph of a crayfish muscle fiber during local contraction initiated by a strong 0.8 ms current pulse producing depolarization of 600 mV. Note that the sarcomere spacing at the center remains unchanged, while contraction is observed at both the upper and the lower edges. Sugi and Ochi (1967a).

13.1.5 Measurement of Velocity of Transverse Spread of Contraction

Figure 139 is a streak photograph of local contraction shown in Fig. 112. The contraction occurs symmetrically with respect to the pipette axis. By changing the slit position with respect to the projected image of the fiber, it was found that the contraction starts at the fiber surface under the pipette and then propagates transversely around the fiber perimeter with a velocity of ~1 cm/s (range, 0.5–3 cm/s, temperature 21–24°C). This propagation velocity of contraction seemed to be consistent with the idea that the transverse spread of contraction was caused by the propagation of the regenerative process along the T tubule network, if the small diameter of T tubules is taken into consideration as well as the low safety factor for propagation of the regenerative process in the T tubule network.

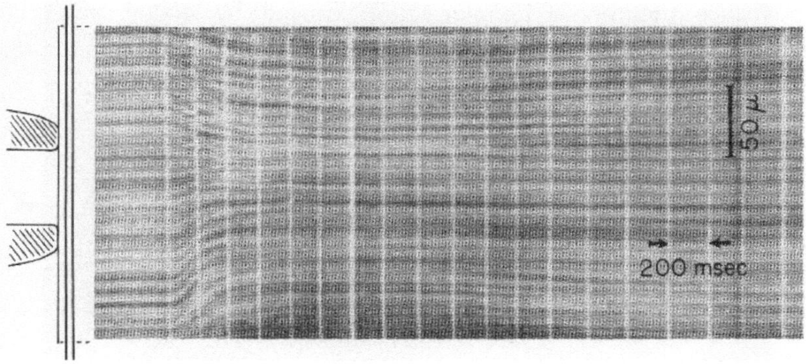

Figure 139 Streak photograph of local contraction spreading along the fiber perimeter, obtained by projecting cinematographic record through a narrow slit to a moving 35 mm film. The position of the pipette and the slit relative to the pipette for local membrane depolarization is shown diagrammatically on the left (Sugi and Ochi, 1967a).

13.1.6 Two Different Types of Spread of Contraction

Figure 140 summarizes the results obtained from crayfish muscle fibers. Both types of local contraction are believed to be caused by the spread of membrane depolarization from the surface membrane to T tubules. With moderate membrane depolarization, which does not cause reduction of membrane resistance under the pipette, the spread of depolarization along the T tubule membrane is graded according to the strength and duration of current pulse applied to the pipette, since the strength of current flowing out of the T tubule membrane is limited to be small.

On the other hand, if strong current pulse applied to the pipette reduces the membrane resistance markedly under the pipette, a pathway opens for strong currents flowing across the fiber membrane and also across the T tubule membrane to set up a regenerative process propagating around the fiber perimeter. The possible role of the regenerating process in crayfish muscle fibers is not clear, since contraction in crayfish muscle fibers in the animal body is known to be graded as controlled by localized junction potentials. In vertebrate skeletal muscle fibers, whose contraction is normally coupled with all-or-none action potential, the regenerative process in the T tubule membrane should perform an essential role in the inward propagation of surface membrane depolarization caused by action potential. This point will be discussed in the next section.

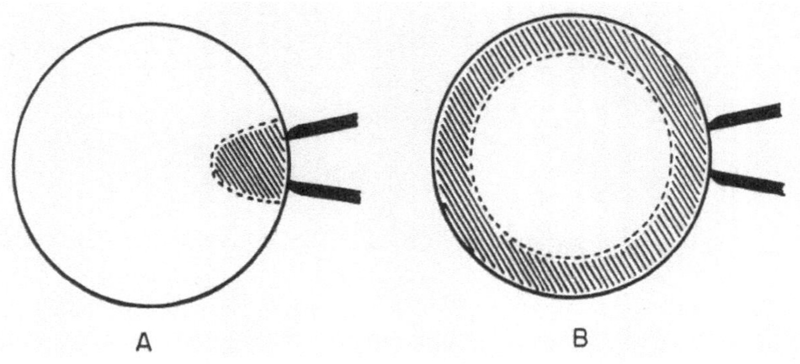

Figure 140 Two different types of local contraction initiated by local membrane depolarization in crayfish muscle fibers. Each circle represents cross-section of the fiber together with pipette in contact with the fiber surface. (A) Graded contraction initiated by moderate membrane depolarization, both the strength and the degree of inward spread of which are dependent on the magnitude and the duration of membrane depolarization. (B) Contraction initiated by strong membrane depolarization, which spreads around the whole perimeter of the fiber (Sugi and Ochi, 1967a).

13.2 Frog Muscle Fibers

13.2.1 General Features of Local Membrane Depolarization

To confirm whether the results obtained by using crayfish muscle fibers can also be obtained by using vertebrate skeletal muscle fibers, we performed experiments on single frog skeletal muscle fibers using the same local activation method. Single muscle fibers isolated from the semitendinosus muscle of the frog (*Rana Japonica*) was used as material. The fiber (diameter, 40–100 μm) was mounted horizontally by clamping both ends. The experimental methods were identical with those of Sugi and Ochi (1967a), except that the pipettes for local activation were 3–40 μm in diameter. The values of contact resistance of the pipette ranged from 100 to 500 kΩ, varying inversely with tip diameter.

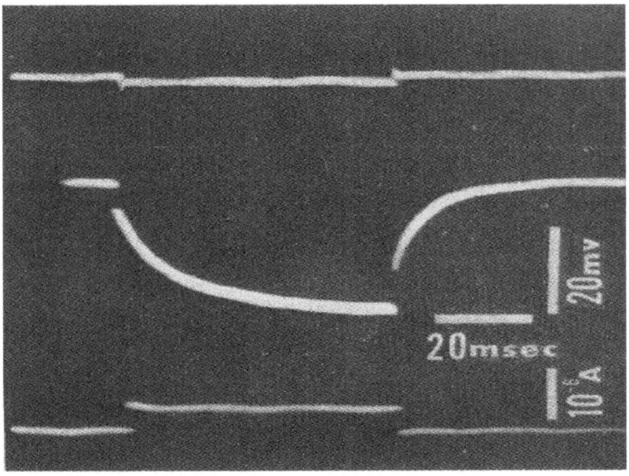

Figure 141 Change in membrane potential of a single muscle fiber during application of a moderate (upper trace) and a strong (middle record) current pulse (lower record), and during application of a strong current pulse (Lower record) (Sugi and Ochi, 1967b).

Figure 141 shows the change in membrane potential of the fiber during the application of negative current pulse to the pipette, recorded with a microelectrode inserted into the fiber near the pipette. No appreciable membrane potential changes were recorded during the application of a moderate current pulse producing membrane depolarization up to 100 mV (upper record). A strong current pulse producing membrane depolarization >100 mV produced irreversible reduction of membrane resistance under the pipette, indicated by the appearance of membrane hyperpolarization around the pipette (middle record), as has been the case in crayfish muscle fibers (Sugi and Ochi, 1967a). As a result of the reduction of membrane resistance under the pipette, strong currents flowed into the fiber around the pipette and flowed out of the fiber at the fiber surface under the pipette, thus producing depolarization across the T tubule membrane located under the pipette.

13.2.2 Inward Spread of Contraction Initiated by Moderate Membrane Depolarization

As shown in Fig. 142, local contraction initiated by a moderate current pulse was localized under the pipette and never spread in longitudinal directions. In agreement with Huxley and Taylor (1958), the extent of the inward spread of contraction was graded, increasing with increasing magnitude and duration of depolarization. The critical membrane depolarization for producing a just perceptible contraction was 20–50 mV for the pipette of 6–40 μm. If a prolonged moderate current pulse was applied to the fiber with a pipette with diameter half as large as the fiber diameter, the contraction spread transversely across the center of the fiber toward the other edge of the fiber. These features of graded local contraction were similar to that of crayfish muscle fibers (Fig. 134–136).

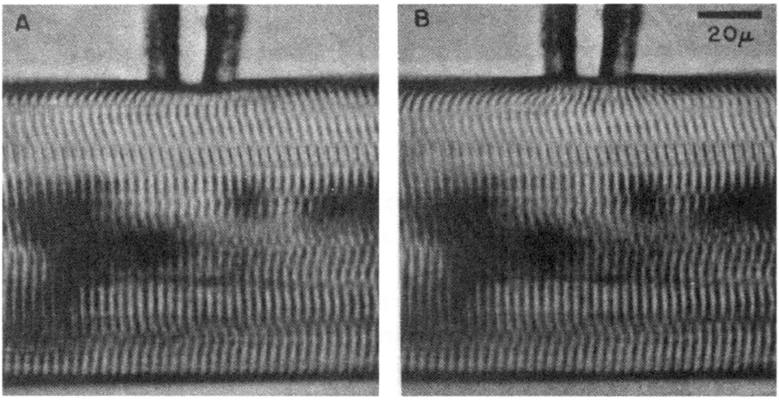

Figure 142 Selected frames from a cinefilm showing a frog muscle fiber before (A) and during (B) application of a 100 ms moderate current pulse producing local membrane depolarization of 50 mV.

13.2.3 Transverse Spread of Contraction Initiated by Strong Current

As with crayfish muscle fibers, a strong current pulse producing membrane depolarization >100 mV initiated the contraction spreading around the whole perimeter of frog muscle fibers. Figure 143 is a typical example showing the nature and reversibility of this type of contraction. It will be seen that the contraction is observed not only at the edge of the fiber under the pipette but also at the opposite edge of the fiber. The reason for this appearance of contraction is that in the photographic records, the fiber surface is not in focus except for the upper and the lower edge of the fiber image; by moving the objective lens up and down to observe fiber surface between the upper and the lower edges, it was confirmed that contraction occurs around the whole fiber perimeter.

Despite marked irreversible reduction of membrane resistance under the pipette, the contraction spreading transversely around the fiber perimeter could be initiated many times without any visible damage to the fiber. The smallest diameter of the pipette with which the contraction spreading around the fiber perimeter could be elicited was ~6 μm. An example of the contraction confined to a few sarcomeres is shown in Fig. 119. This type of contraction could still be initiated by replacing Na^+ in the experimental solution by choline, indicating that the regenerative process in the T tubule membrane causing the contraction is not the action potential associated with the inward movement of Na^+.

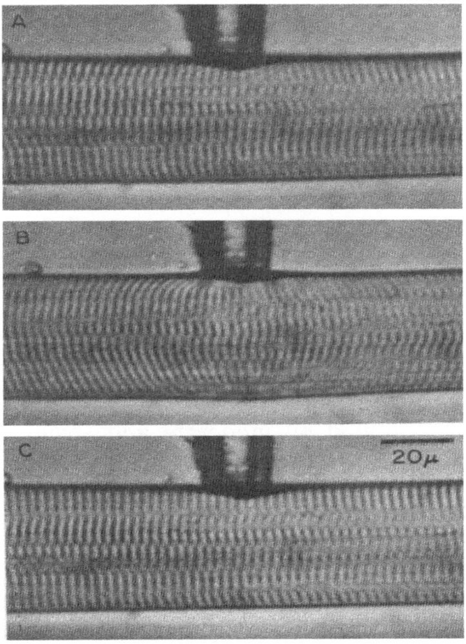

Figure 143 Selected frames from a cinefilm showing a muscle fiber during (A and B), and after (C) application of a 100 ms strong current pulse producing depolarization of 150 mV. Note that the contraction is observed at both the upper and the lower edge of the fiber (Sugi and Ochi, 1967b).

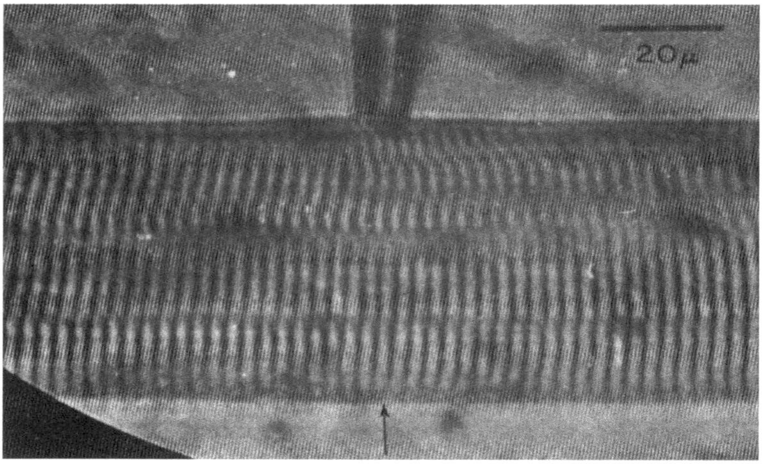

Figure 144 Photomicrograph showing contraction spreading around the fiber perimeter, which is confined to a few sarcomeres. The contraction at the lower edge is indicated by an arrow. Sugi and Ochi (1967b).

13.2.4 Measurement of Velocity of Transverse Spread of Contraction

Figure 145 shows examples of streak photographs of local contraction initiated by a strong current pulse to determine the velocity of propagation of contraction around the fiber perimeter. The value of propagation velocity ranged from 0.8 to 6 cm/s (24–26°C). This value seems to be compatible with the reports that the diameter of T tubules is ~0.03 μm (Peachey, 1965). These results indicate that there is no qualitative difference between frog and crayfish muscle fibers in the two different modes of contraction initiated by moderate and strong current pulses. The reason why contraction spreads around the fiber perimeter but not across the center of the fiber is unknown; probably the T tubule network near the fiber surface may provide favorable conditions for the propagation of regenerating process than does the T tubule network at the center of the fiber. Complex nature of the T tubule network has been indicated by these results and also by computer calculations reported by many authors that when the limited area of the muscle fiber surface is moderately depolarized by the voltage-clamp methods, the resulting contraction is most marked at the area distant from the fiber surface, but not at the area nearest the fiber surface (for a review, see Nakajima and Bastian, 1976).

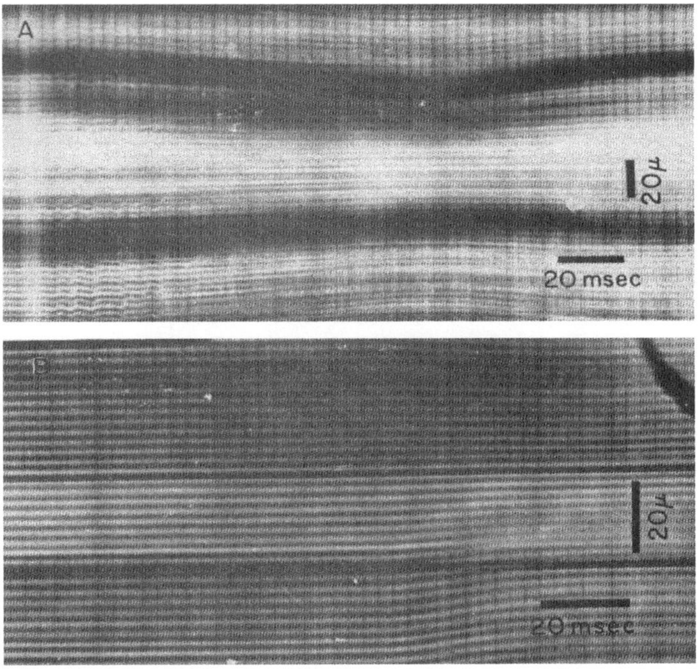

Figure 145 Streak photographs of local contraction spreading around the fiber perimeter, obtained by the method identical with that used previously. The onset of current pulse is indicated by a flash of light from the xenon lamp in A (vertical white line at the beginning of the record), and by upward deflection of electromagnet in B (black shadow at the top of the record. In both A and B, parallel black and white lines are cross-striation of the fiber (Sugi and Ochi, 1967b).

When Dr. Ochi and I completed local activation experiments on crayfish and frog muscle fibers, I had an opportunity to report our results in front of Professor A. F. Huxley on the occasion of the XXII International Congress of Physiological Sciences held in Tokyo in 1995. After the session, he came to me and said, *"I am very much interested in your work. If possible, I would like to visit your laboratory to discuss the results you reported."* Next day, he came to my laboratory and discussed possible regenerative process in the T tubule network (Fig. 121). Since then, I had very pleasant terms with him over almost half a century.

Figure 146 Photograph taken in 1965 at the Department of Physiology, University of Tokyo. From left to right, Dr. Michihiko Ogata (Kyushu University), Dr. Richard J. Podolsky (National Institutes of Health), Professor Koji Uchizono (University of Tokyo), I, and Professor Andrew F. Huxley (Cambridge University at that time).

13.3 Accommodation of E–C Coupling Process as Studied by Local Membrane Depolarization with Linearly Rising Currents

As early as 1907, Lucas reported that if a nerve fiber was stimulated with linearly increasing current, there was a minimal current gradient to evoke an action potential; currents with less steep gradient did not evoke action potential even when the resulting membrane potential far exceeded the threshold value for rectangular current pulses. This phenomenon is called "accommodation" and is studied on a number of different excitable cells and tissues. Up to the mid-1960s, however, little information was available on the accommodation process in excitation–contraction coupling, i.e., the coupling between surface membrane depolarization and Ca^{2+} release from the SR.

In both summation of force responses (Fig. 106) and summation of local contraction (Fig. 111), successive short current pulses were found to be much more effective than prolonged current pulses in producing large summated responses. This indicates that the effect of membrane depolarization in producing contraction decreases with time during the application of a prolonged depolarization.

As a matter of fact, it has been known that K-contractures of muscle fibers in high-K^+ medium relax spontaneously, indicating that the effect of membrane depolarization in producing contraction decreases with time in high-K^+ medium to eventually disappear. Hodgkin and Horowicz (1960) explained this phenomenon as follows: (1) Membrane depolarization releases an activator initiating contraction, and (2) accumulation of its end product causes relaxation of K-contractures. The activator was later proved to be Ca^{2+} released from the sarcoplasmic reticulum (SR) (Winegrad, 1968; Ebashi and Endo, 1968). According to the current knowledge of excitation–contraction coupling mechanism (Ebashi and Endo, 1968), the activator corresponds to Ca^{2+} released from the SR, while the end product corresponds to Ca^{2+} taken up again by the SR (Winegrad, 1968).

When I went to New York at the end of 1965 to start my research work at Columbia University, I decided to study the accommodation process taking place in the T tubule and the SR membranes by applying linearly rising (or triangular) currents to the pipette (diameter, 10–40 μm) in contact with the surface membrane of frog muscle fibers. At the start of experiments, I examined the critical value of local membrane depolarization to cause irreversible reduction of membrane resistance.

13.3.1 Membrane Potential Changes during Application of Linearly Rising Currents for Local Activation

Figure 147 shows membrane potential changes of a single frog muscle fiber (upper records), measured by an intracellular microelectrode inserted into the fiber at the point close to the pipette for local activation, to apply linearly rising currents (duration, 400 ms, lower records). No appreciable change in membrane potential was observed during the application of linearly rising current to the pipette, producing peak membrane depolarization up to 300 mV (Fig. 147A). This indicated that linearly rising currents produce reduction of membrane resistance under the pipette much less readily than do rectangular current pulses. Repeated application of local membrane depolarization, however, eventually produced reduction of membrane resistance under the pipette, as indicated by membrane hyperpolarization during application of linearly rising currents (Fig. 147B, Sugi and Ochi, 1967a,b). In the following experiments on the accommodation of the contractile mechanism, the magnitude of the local membrane never exceeded 100 mV, so that experiments were conducted without causing reduction of membrane potential under the pipette.

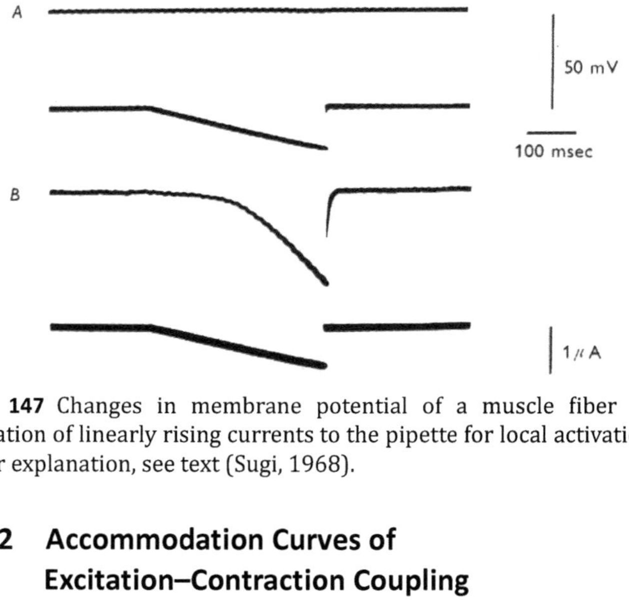

Figure 147 Changes in membrane potential of a muscle fiber during application of linearly rising currents to the pipette for local activation. For further explanation, see text (Sugi, 1968).

13.3.2 Accommodation Curves of Excitation–Contraction Coupling

As shown in Fig. 148, the magnitude of threshold depolarization to initiate just perceptible local contraction under the pipette remained constant over a wide range of duration of linearly rising current from 0.5 to more than 20 s. Similar results were obtained when Na^+ in the experimental solution was replaced with choline, except that the threshold depolarization at the flat part of the accommodation curve increased by ~20%. These results indicate that the accommodation process in the excitation–contraction coupling is normally very slow. The minimal gradient of local membrane depolarization for initiating contraction was ~1 mV/s (16–24°C).

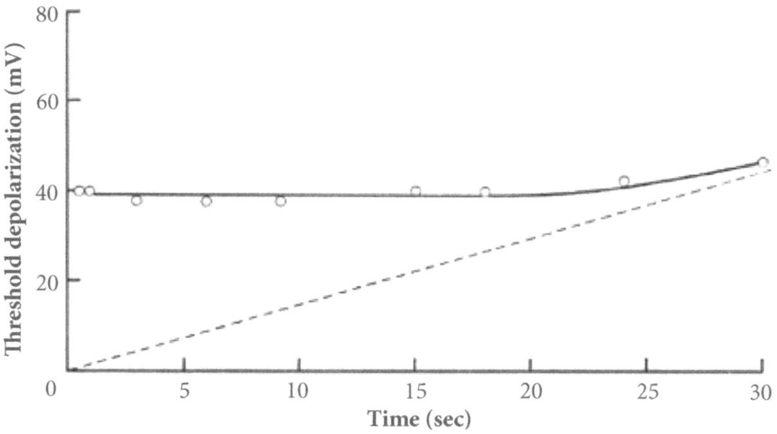

Figure 148 Accommodation curve of excitation–contraction coupling mechanism. In standard experimental condition with 1.8 mM Ca^{2+}. Magnitude of threshold Local membrane depolarization to initiate a just perceptible contraction is plotted. Against duration of linearly rising current. Broken line indicates minimal gradient of depolarization for local activation of a muscle fiber (Sugi, 1968).

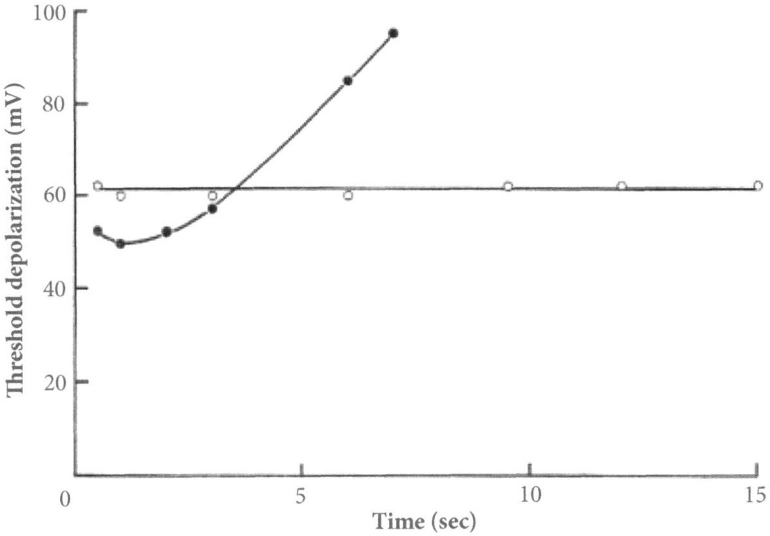

Figure 149 Effect of decreasing [Ca^{2+}] on the rate of development of the accommodation process. Open and filled circles indicate accommodation curve with 1.8 mM Ca^{2+} and 0.1 mM Ca^{2+}, respectively. The curves were obtained in experimental solution, in which Na^+ was replaced by choline (Sugi, 1968).

The rate of the development of the accommodation process increased markedly when [Ca^{2+}] in the experimental solution was decreased from 1.8 to 0.1 mM. As can be seen in Fig. 149, the magnitude of threshold depolarization increased steeply with increasing duration of linearly rising current exceeded ~1 s.

13.3.3 Time Course of Development and Decay of Accommodation Process

Time course of development and decay of the accommodation process was examined by first applying a series of "test" subthreshold linearly rising current of a constant gradient and varying duration, and the threshold membrane depolarization for initiating contraction under the pipette was determined by applying "test" rectangular current pulse of 100 ms duration at the end of each "conditioning" linearly rising current. Figure 150 is an example of results with 0.1 mM Ca^{2+}. Threshold depolarization increased exponentially with increasing duration of "conditioning" subthreshold linearly rising current. Time course of decay of the accommodation process, i.e., time course of recovery of threshold depolarization, was also examined

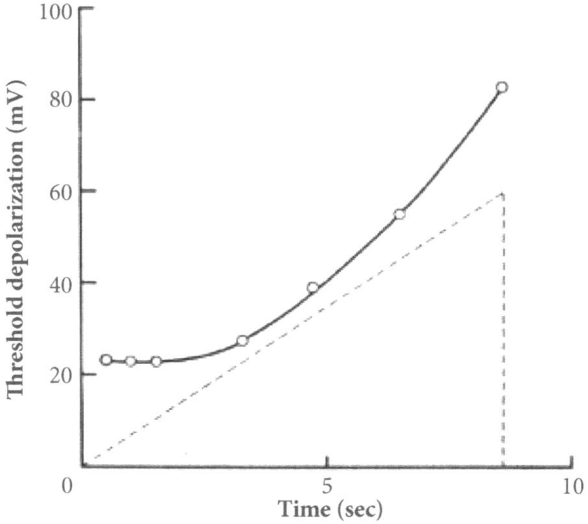

Figure 150 Time course of the accommodative rise of threshold depolarization for 100 ms rectangular current pulse during application of "conditioning" subthreshold linearly rising current (broken line) (Sugi, 1968).

by applying "test" rectangular 100 ms current pulses at various times after termination of "conditioning" subthreshold linearly rising current. As can be seen in Fig. 151, the recovery of threshold depolarization for contraction also decreased exponentially after termination of "conditioning" linearly rising current.

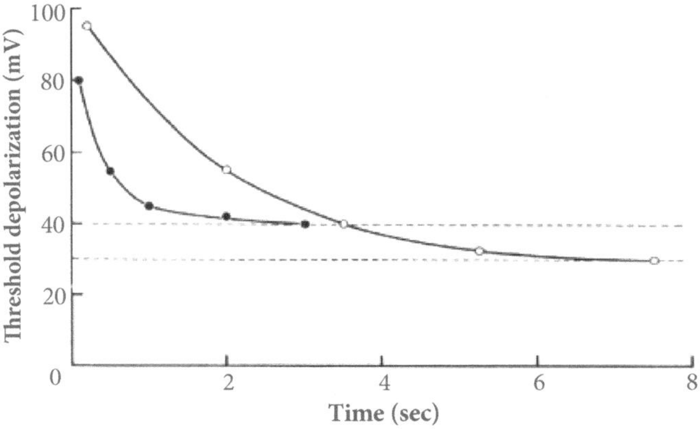

Figure 151 Time course of decay of the accommodation process after termination of "conditioning" subthreshold linearly rising current. The curves were obtained from two different fibers in experimental solution with 0.1 mM Ca^{2+}. Duration of linearly rising current was 3 s (filled circles) and 10 s (open circles). Sugi (1968).

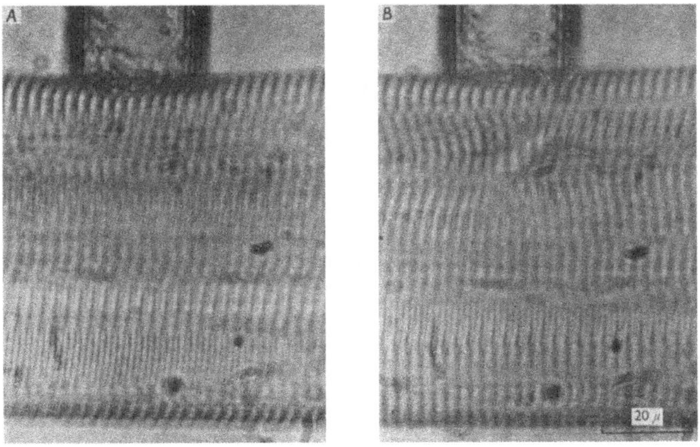

Figure 152 Anomalous local contraction initiated by a 3 s linearly rising current, observed in fibers kept in experimental solution with 0.1 mM Ca^{2+}. Note that contraction occurs only in the region distant from the fiber surface (upper edge of the fiber) where the pipette for local activation is placed. Sugi (1968).

In the experimental solution with 0.1 mM Ca^{2+}, linearly rising current of >3 s duration sometimes elicited local contraction at the region distant from the fiber surface, but not at the fiber surface where the pipette for local membrane depolarization was placed (Fig. 152). With linearly rising current of shorter duration, however, initialed contraction at the fiber surface under the pipette. These observations indicate that the rise of threshold depolarization to initiate contraction as a result of accommodation is most marked at the most superficial layer of the fiber and less marked with increasing distance from the fiber surface.

13.3.4 Accommodation of Action Potential in Muscle Fiber

The accommodation process in the generation of action potential in muscle fibers was also examined by applying linearly rising currents to a glass microelectrode inserted into the fiber and recording the resulting generation of action potential with another microelectrode inserted into the fiber at the point 50–100 μm distant from the current electrode. The threshold depolarization for action potential generation was 20–40 mV and remained fairly constant for linearly rising currents up to 0.5–2 s duration (Fig. 153). With longer linearly rising currents, local contraction was initiated without generation of action potential, indicating the accommodation process of action potential generation proceeds more rapidly than that of excitation–contraction coupling.

13.4 Inward and Transverse Spread of Contraction Studied Using High-Speed Cinematography

In 1973, I was appointed professor and chairman in the Department of Physiology, Teikyo University Medical School and could purchase expensive equipment, such as a 16 mm high-speed cine-camera (Hicam, Redlake Corp.), with which local contractions can be recorded at 1,000–3,000 frames/s, enabling me to study modes of inward spread of contraction in frog muscle fibers in detail under physiological conditions. By the method of high-speed cinematography, together with the great number of muscle fibers studied, I could find many new features of local contraction.

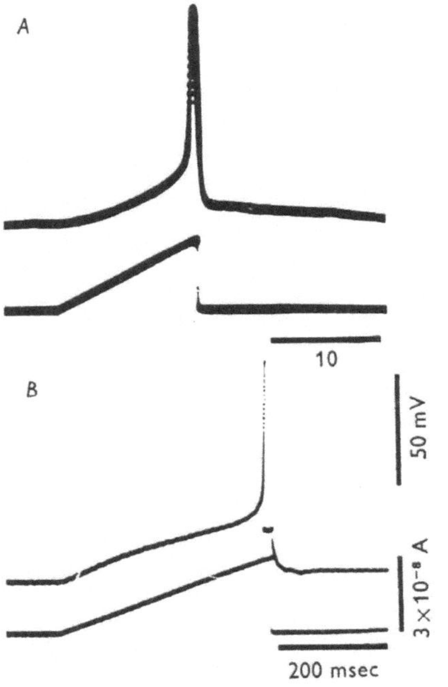

Figure 153 Records of muscle fiber action potential (upper traces) elicited by intracellularly applied linearly rising current Lower traces). Note that the threshold (or critical) depolarization for action potential generation is nearly the same between short (A) and long (B) linearly rising currents (Sugi, 1968).

The experimental work to be described in this section was my last work concerned with the research field of excitation–contraction coupling.

13.4.1 General Features of Phasic-Type Local Contraction Initiated by Moderate Membrane Depolarization

All experiments were performed with single fibers isolated from the semitendinosus muscle of the frog *Rana nigromaculata* and

Rana japonica. It was found that there are two different types of muscle fibers; one shows only graded local contraction, while the other exhibits phasic local contraction in response to moderate local membrane depolarization (20–30 mV) applied with pipettes of 30–40 μm diameter. The phasic-type local contraction was observed in ~30% of the fibers studied.

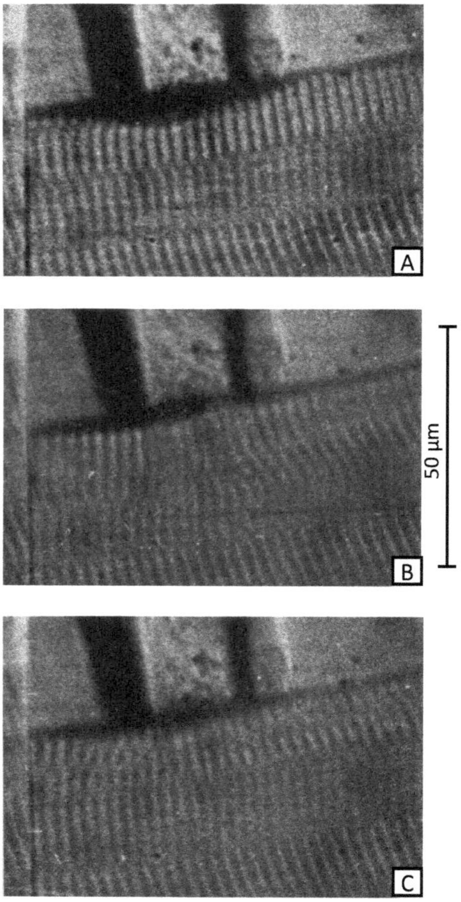

Figure 154 Selected frames from a high-speed cinefilm (3,000 frames/s) showing a phasic-type local contraction elicited by a 500 ms rectangular current pulse producing local membrane depolarization of 25 mV. (A) Resting fiber. (B) A frame taken at 50 ms after the onset of current pulse. (C) A frame taken at 150 ms after the onset of current pulse. Note that the local contraction under the pipette starts to relax while membrane depolarization is still maintained (Sugi, 1974).

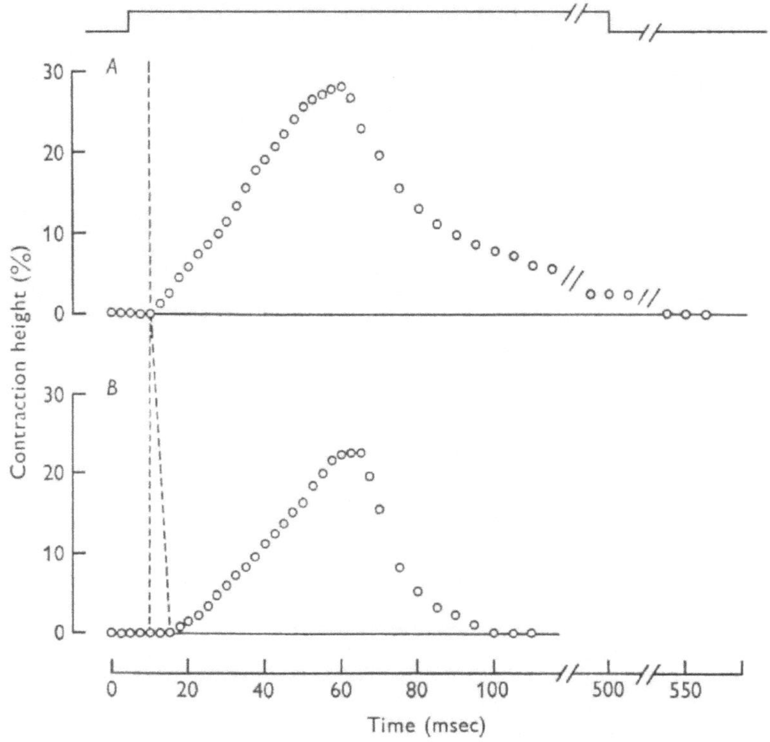

Figure 155 Time course of the local contraction shown in Fig. 154, obtained by measuring the length of five adjacent sarcomeres at the depolarized fiber surface under the pipette (A), and at the inner region 14 μm distant from the fiber surface (B). In this and subsequent figures, contraction height represents the degree of sarcomere shortening in percentage of the initial length. In this and subsequent figures, time course of current pulse producing local membrane depolarization is shown at the top. Broken lines indicate the difference in the time of onset of contraction between the two regions. Sugi (1974).

A typical high-speed cinematography showing the phasic nature of the contraction is presented in Fig. 154. When the fiber was locally activated by a pipette in contact with the fiber surface (A), local contraction was initiated under the pipette (B), which relaxed completely while the current pulse producing local membrane depolarization still continued (C). Figure 155 shows contraction curves obtained from the cinematographic records by measuring the length of five consecutive sarcomeres at the depolarized fiber surface (A), and at the region 14 μm distant from the fiber surface (B). It can be seen that (1) though the contraction starts to relax ~60 ms after the onset of contraction, small residual contraction remained at the depolarized fiber surface as long as depolarization continued; (2) on the other hand, the contraction at the region distant from the fiber surface relaxed completely after ~100 ms after onset of contraction; and (3) from the difference in time of onset of contraction between the two different regions, the contraction was found to spread inward with a velocity of 0.7–2 cm (18–26°C). These findings are consistent with the idea that some regenerative process is involved in the inward spread of contraction in the phasic type of local contraction.

13.4.2 Effect of Increase in Area of Depolarization

When the fiber was locally depolarized for 300–500 ms with a large pipette of diameter 50–60 μm, the phasic-type contraction exhibited a conspicuous nature. As shown in Fig. 156, the contraction was first initiated at the region ~15 μm distant from the depolarized fiber surface (A) and then spread not only inward but also outward to the depolarized fiber surface (B, C). Contraction curves in five adjacent sarcomeres (Fig. 157) indicated that (1) the peak contraction height was larger at the region distant from the fiber surface and (2) the contraction at the inner region persisted as long as the depolarizing current continued, while the contraction at the fiber surface relaxed completely in ~100 ms. The velocity of outward spread of contraction was ~0.3 cm/s.

These features of the phasic-type contraction initiated by a large area of membrane depolarization may be taken to indicate that the T-tubule network structure is such that, if a large area of fiber surface is depolarized, the resulting depolarization across the T tubule membrane is largest at a region distant from the fiber surface. This explanation is consistent with the observation that if part of the muscle fiber surface membrane is depolarized by the voltage clamp method, contraction first occurs at sarcomeres distant from the fiber surface (Costantin, 1970). This property of T-tubule network may somehow be related to the transverse spread of contraction around the whole fiber perimeter observed by Sugi and Ochi (1967a,b).

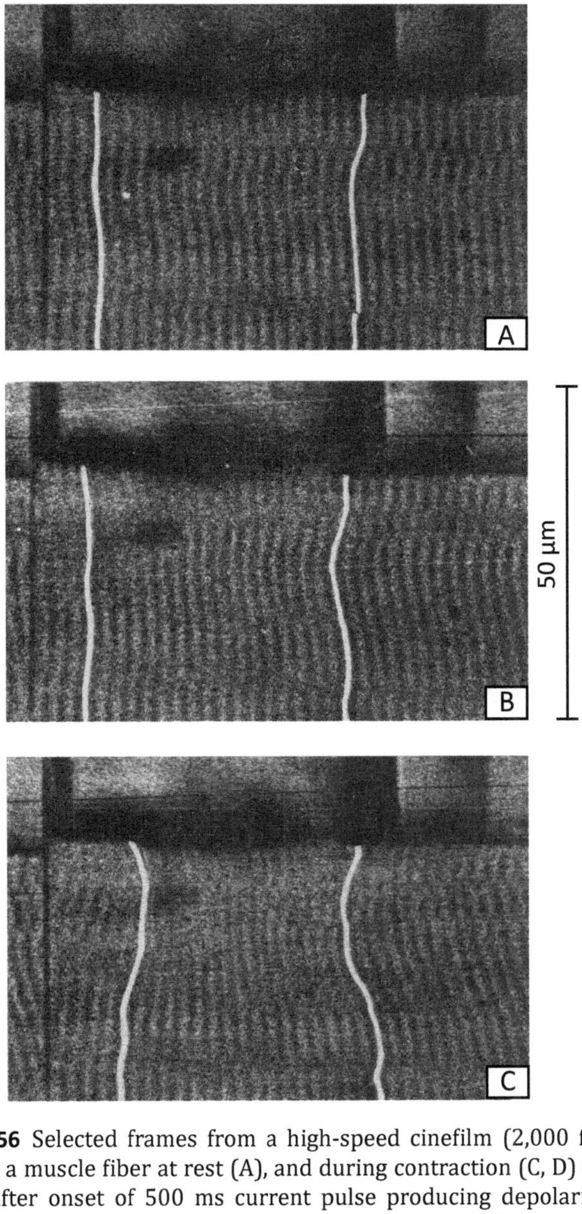

Figure 156 Selected frames from a high-speed cinefilm (2,000 frames/s) showing a muscle fiber at rest (A), and during contraction (C, D) at 40 and 40 ms after onset of 500 ms current pulse producing depolarization of 25 mV, respectively. White lines in each frame mark eighteen sarcomeres under the pipette. Note that the contraction is first initiated at the region 15 μm distant from the fiber surface, and the extent of contraction is most marked at this inner region during the course of contraction (Sugi, 1968).

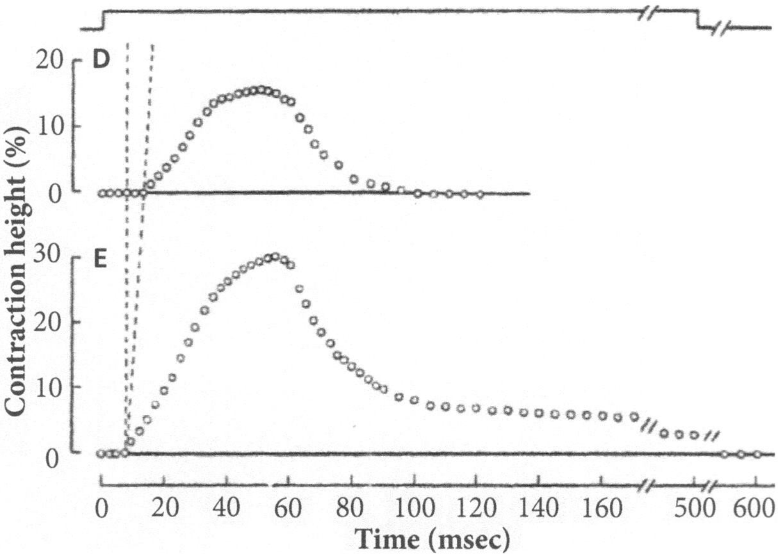

Figure 157 Contraction curves of the local contraction shown in Fig. 156. The curves were obtained by measuring ten adjacent sarcomeres at the depolarized fiber surface (D) and at the region 15 μm distant from the fiber surface (E). Note that peak contraction height is much larger at the inner region distant from the fiber surface (E) than at the fiber surface (D), as well as the outward spread of contraction from the inner region to the fiber surface (Sugi, 1968).

13.4.3 Effect of Magnitude of Depolarization

Figure 158 shows contraction curves of phasic-type local contraction at the fiber surface and at the region 16 μm distant from the fiber surface, elicited by 500 ms current pulses applied to a large pipette (diameter, 65 μm) producing depolarization of 20 mV (A, B) and 30 mV (C, D). In response to 20 mV depolarization, relaxation of contraction started at ~40 ms after the onset of current pulse, and was complete at the fiber surface in ~100 ms after the onset of current pulse (A), while the residual contraction lasting as long as depolarization continued was seen at the inner region of the fiber (B). In response to 30 mV depolarization, on the other hand, the contraction exhibited a prolonged plateau, so that relaxation started ~70 ms after the onset of current pulse and proceeded with a slow rate (C,D); at both the fiber surface and at the inner region, the residual contraction lasted as long as current pulse went on. These results suggest that as the magnitude of depolarization is increased, the resulting contraction consisted of both phasic and graded type of contractions.

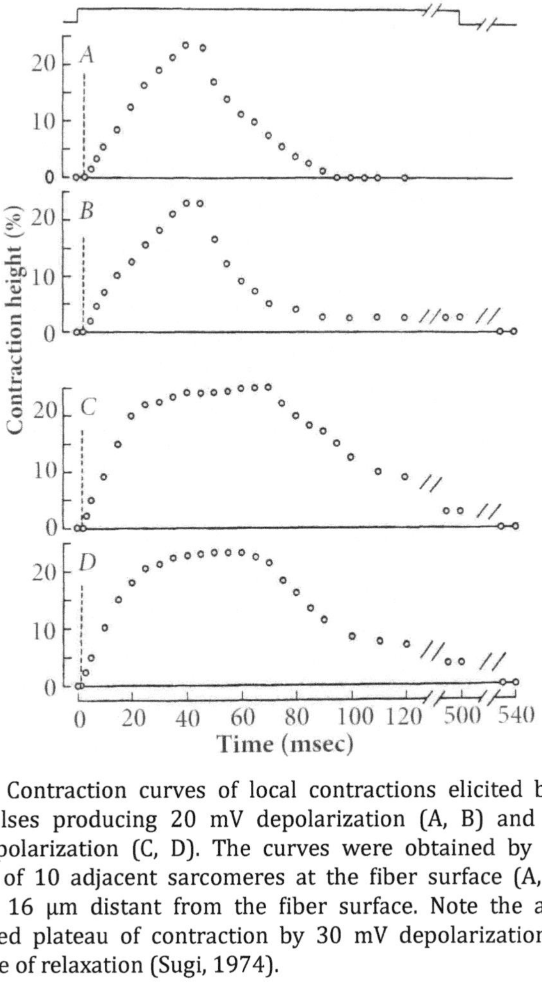

Figure 158 Contraction curves of local contractions elicited by 500 ms current pulses producing 20 mV depolarization (A, B) and producing 30 mV depolarization (C, D). The curves were obtained by measuring the length of 10 adjacent sarcomeres at the fiber surface (A, C) and at the region 16 μm distant from the fiber surface. Note the appearance of prolonged plateau of contraction by 30 mV depolarization, followed by slow rate of relaxation (Sugi, 1974).

13.4.4 Transverse Spread of Contraction Across the Center to the Opposite Side of the Fiber

In some fibers, the phasic-type local contraction was found to spread not only inward but also further across the center to the opposite side of the fiber. As can be seen in Fig. 159, in a few sarcomeres under the pipette (diameter, ~50 μm), the contraction spreads not only inward but also transversely up to the lower edge of the fiber, indicating that the contraction spreads transversely all the way across the fiber. Such contraction spreading across the fiber could only be initiated by large pipettes.

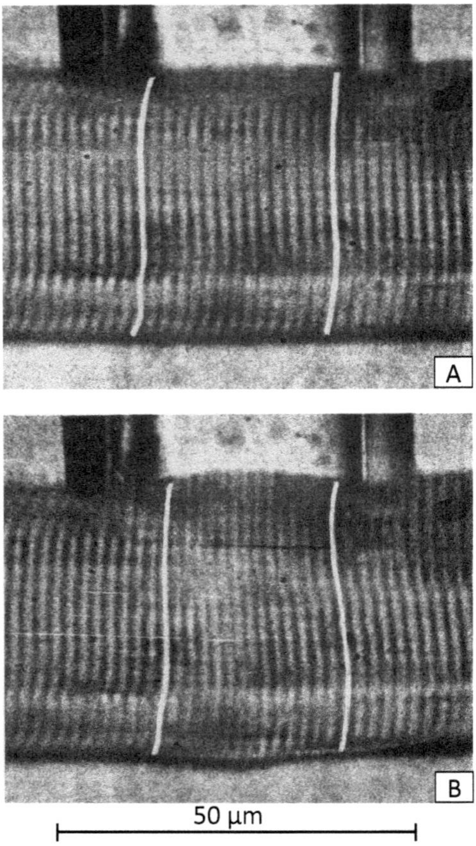

Figure 159 Selected frames from a high-speed cinefilm (2,000 frames/s) showing the phasic-type local contraction spreading transversely across the center to the opposite side of the fiber. (A) Resting fiber. (B) During contraction elicited by a 500 ms current pulse producing membrane depolarization of 25 mV. White lines mark 15 sarcomeres under the pipette. Note that the contraction spreads across the fiber with respect to a few sarcomeres under the pipette (diameter, ~50 μm). Sugi (1974).

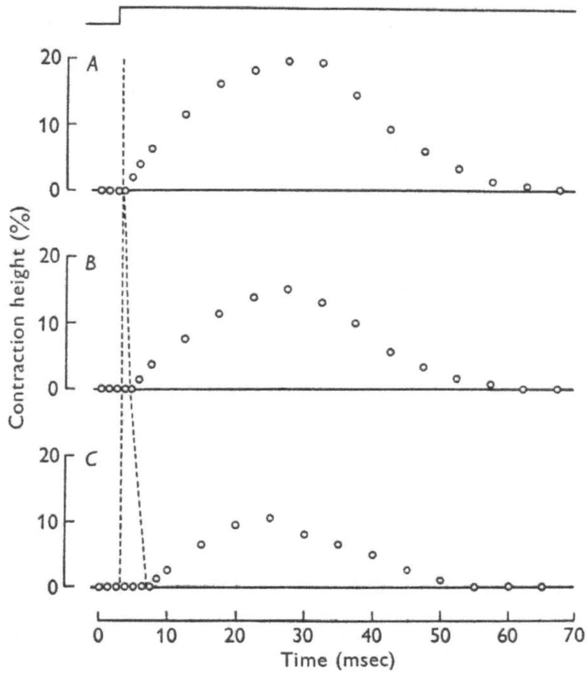

Figure 160 Contraction curves of phasic-type local contraction shown in Fig. 159, obtained by measuring the length of five adjacent sarcomeres at the fiber surface under the pipette (A), at the inner region 20 μm (B) and 38 μm distant from the fiber surface (C). Sugi (1974).

As can be seen in Fig. 160, the contraction relaxed completely in ~60 ms during the application of a 500 ms current pulse producing depolarization of 25 mV. The peak contraction height decreased with increasing distance from the depolarized fiber surface, indicating a low safety factor for propagation of the regenerating process in the T tubule network. The velocity of spread of contraction across the fiber was ~1 cm/s and appeared to decrease with increasing distance from the fiber surface.

13.4.5 Effect of Temperature

Figure 161 shows early part of contraction curves of the phasic-type local contraction elicited by a 300 ms current pulse producing depolarization of 25 mV, obtained at 13.5°C (A,B) and at 23.5°C (C,D). Both the latent period, i.e., the time between the onset of current pulse and that of contraction, and the velocity of spread of contraction decreased with decreasing temperature. From other experiments, in which temperature of experimental solution was varied from 5 to 25°C, the value of Q_{10} was estimated to be ~2.3 for the velocity of spread of contraction, and ~2.8 for the latent period of contraction. Since electrical parameters of T tubules network is unlikely may not change appreciably in the above range of temperature, the high Q_{10} values may be associated with the process linking the T tubule membrane depolarization to Ca^{2+} release from the SR.

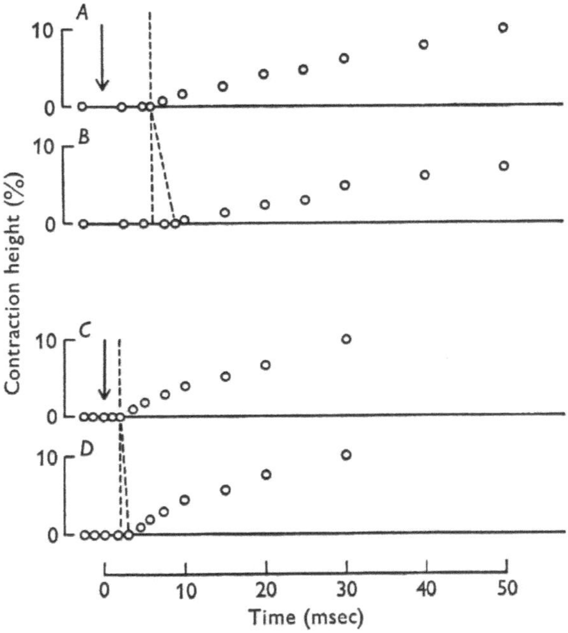

Figure 161 Early part of contraction curves of phasic-type local contraction, initiated membrane depolarization of 25 mV with a pipette of 40 μm diameter. The curves are obtained at the depolarized fiber surface (A, C) and at the region 15 μm distant from the fiber surface (B, D). Curves A, B and C, D were obtained at 13.5 and 23.5°C, respectively. Downward arrows indicate time of onset of current pulse. Sugi (1974).

13.4.6 Shortening Velocity of Sarcomeres

In fibers with initial sarcomere length of 2.5–3 μm, the shortening of sarcomeres occurred almost linearly with time for the first three quarters of sarcomere shortening (Figs. 155, 157 and 158). The phase of constant velocity sarcomere shortening became less marked if initial sarcomere length was less than 2 μm (Fig. 161). The constant sarcomere shortening velocity per sarcomere was 15 to 25 μm/s at 18–26°C, comparable to the velocity of steady shortening in tetanized muscle fibers under a load of 0.05 P_0 (~9.1 L_0/s or ~20 μm/sarcomere at 26°C, Buchthal and Kaiser, 1951, p175). The steady sarcomere shortening velocity was reduced to 3–5 μm/s at 5–6°C. This value is also comparable to the steady shortening velocity of muscle fibers under a very small load at 6°C (Gordon, Huxley and Julian, 1966).

These results indicate that (1) sarcomeres can be fully activated by local membrane depolarization and (2) the fully activated sarcomeres shorten against a very small load. Point (2) can readily be understood because the activated sarcomeres shorten by stretching adjacent sarcomere in relaxed state.

13.4.7 Graded-Type Local Contraction

The experiments were also made on the fibers, which showed only graded local contractions in response to local membrane depolarization. When 100–500 ms current pulses producing depolarization of 20–30 mV were applied to pipettes of diameter less than 40 μm, the contraction was first initiated at the depolarized fiber surface and spread inward with a velocity of 0.6–1.5 cm/s in a graded manner depending on the magnitude and duration of depolarization (Sugi and Ochi, 1967b; Sugi, 1968).

In contrast, when the fiber was locally depolarized by applying current pulses to pipettes of diameter 50–60 μm, the contraction was larger in the inner region than at the depolarized fiber surface, as can be seen in Fig. 162. This feature of graded local contraction is qualitatively similar to that of phasic-type local contraction initiated by large pipettes, shown in Fig. 156, again indicating that if a large area of fiber surface is depolarized, the magnitude of resulting depolarization of the T tubule membrane may be larger at the inner region than at the fiber surface. In both regions, the contraction continued as long as membrane depolarization continued.

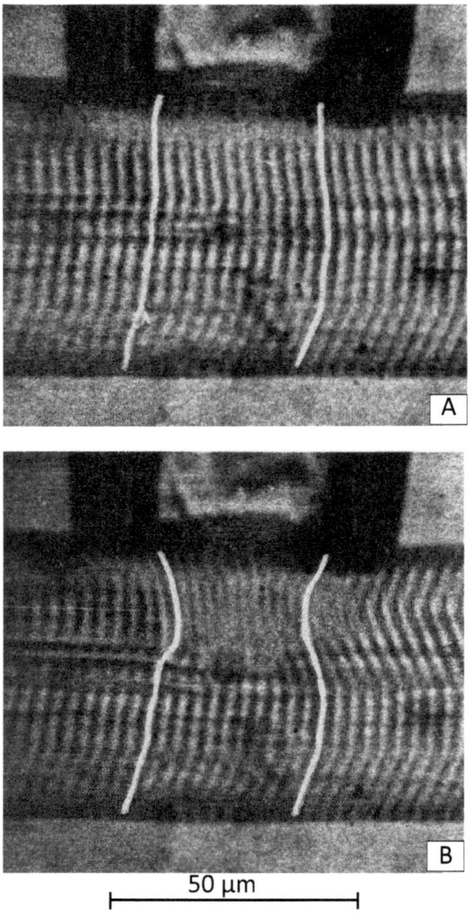

Figure 162 Selected frames from a high-speed cinefilm (2,000 frames/s) showing graded local contraction produced by a 300 ms current pulse producing depolarization of 80 mV. Pipette diameter, ~60 µm. White lines mark 12 sarcomeres under the pipette. Note that the contraction is most marked at the region distant from the fiber surface (Sugi, 1974).

In Fig. 163, contraction curves obtained from the cinematographic record shown in Fig. 162 are presented. It can be seen that the contraction first takes place at the inner region and spreads back to the depolarized fiber surface. This feature of graded local contraction was also similar to that observed in phasic-type local contraction initiated by depolarization over a large area of fiber surface (Fig. 151).

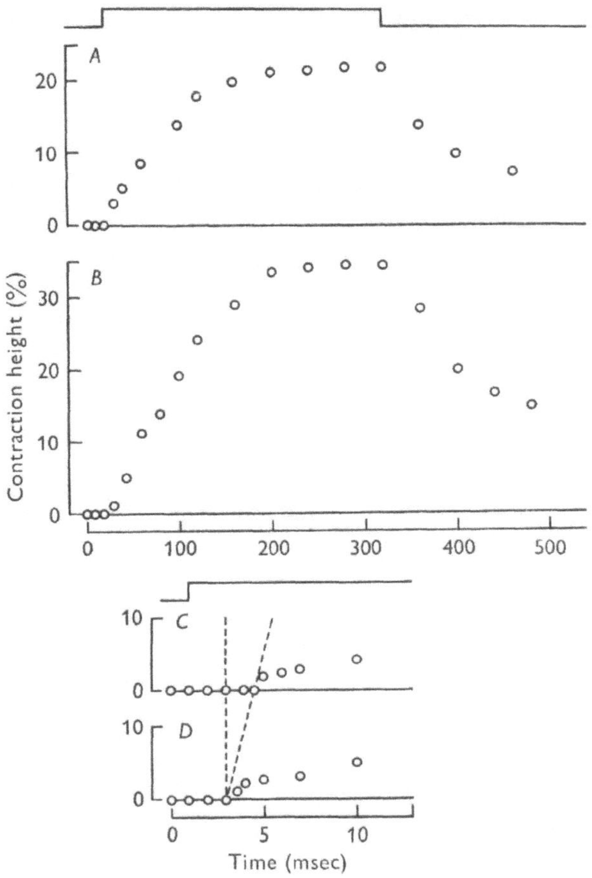

Figure 163 Contraction curves of graded local contraction by measuring length of five adjacent sarcomeres in the cinematographic records shown in Fig 162. (A and C) at the depolarized fiber surface. (B and D) at the region 15 μm distant from the fiber surface. Note that the contraction starts to relax after the end of membrane depolarization (Sugi, 1974).

The velocity of inward (and outward) spread of graded local contraction, measured by the difference in time of onset of contraction after onset of depolarization, was 0.6–1.5 cm/s in accordance with the previous report (Sugi and Ochi, 1967b). On the other hand, the velocity of sarcomere shortening in graded local contraction was less than 10 μm/s, much slower than that in phasic-type local contraction.

13.4.8 Possible Presence and Role of Regenerative Process in Muscle Contraction

At the end of this chapter, I will summarize evidence for the presence of regenerative process in the T tubules of vertebrate skeletal muscle fibers, mainly obtained by my high-speed cinematographic studies on the phasic-type local contraction. The phasic contraction has the following characteristics: (1) It has a definite threshold depolarization of 20–30 mV. (2) It relaxes spontaneously while depolarization still continues (Figs. 155, 157, 158, 160). (3) It spreads inward or transversely across the fiber with a velocity ranging from 0.7 to 2 cm/s. (4) The value of Q_{10} for the velocity of spread of contraction is about 2.3. (5) It is associated with sarcomere shortening with constant velocity, which is comparable to the velocity of isotonic shortening of muscle fibers under very small loads.

These characteristics of phasic-type local contraction indicate that at least in muscle fibers exhibiting the phasic-type local contraction, the T tubules have some kind of regenerative process, which contributes to the spread of surface membrane depolarization along the T tubule membrane into the interior of the fiber. On this basis, the phasic local contraction can be understood by assuming that brief transient potential changes analogous to muscle fiber action potential are set up in the T tubule membrane to produce rapid phasic contractions shown in Figs. 155 and 157.

If the area of surface membrane depolarization was increased by using pipettes of large diameter, the contraction was first initiated at the region distant from the fiber surface (Fig. 157). A similar observation has been made by Costantin (1970), who depolarized the muscle fiber membrane by the voltage clamp method. This can be accounted for by assuming that the density of the current flowing across the T tubules membrane is larger at the fiber inner region than at the fiber surface, so that the regenerative process starts first at the inner region. This property may somehow be related to the spread of contraction around the whole fiber perimeter observed by Sugi and Ochi (1967a,b).

When the magnitude of membrane depolarization was increased, the phasic nature of contraction became less conspicuous; it showed a prolonged plateau followed by slow spontaneous relaxation, and more marked residual contraction lasted until the end of current pulse producing depolarization (Fig. 158). This can be due to the increasing contribution with time of graded contraction, which is produced by passive electrotonic spread of surface membrane depolarization along the T tubule membrane.

As already discussed, the sarcomeres under the pipette for local activation are fully activated like tetanized muscle fibers, indicating that the role of regenerating process is to rapidly and fully activate muscle fibers by a train of action potentials. In this connection, it is of interest that there is a large variation in the twitch-tetanus ratio in both single fibers (Ramsey and Street, 1941) and whole muscles (Close, 1972). If the large twitch is caused by the regenerative process, the twitch-tetanus ratio in whole muscles may reflect the proportion of fibers having the regenerative process in the T tubules.

Chapter 14

Epilogue

During the course of the writing of this book, I realized how good luck struck me so many times, so that I got opportunities to make wide-ranging contributions to the research field of muscle contraction in my research career over more than 50 years. As described in Chapters 12 and 13, I started my research work in 1962, when I was appointed instructor in Physiology at the University of Tokyo Medical School. At that time, I had the experience to perform experiments as a postgraduate student in the Department of Zoology at the University of Tokyo, where I studied contraction of a protozoan animal, *Carchesium plypinum,* having a long contractile stalk ~1 mm long, generally called the stalk muscle (Fig. 164). I struggled with this small animal under a light microscope over five years and succeeded in measuring an extremely rapid propagation velocity of contraction (20–50 cm/s, ~500 length/s considering its length of ~1 mm) by means of streak photography. In 1962, I graduated from the postgraduate school with a PhD, and moved to the University of Tokyo Medical School. This was my first good luck, because I was already accustomed to performing experiments under microscopic observation of small specimen, and my idea based on my experience on protozoan animals could be unique among physiologists in Japan.

Mysteries in Muscle Contraction: Evidence against Current Dogmas
Haruo Sugi

ISBN 978-981-4774-48-2 (Hardcover), 978-1-315-10191-0 (eBook)
www.panstanford.com

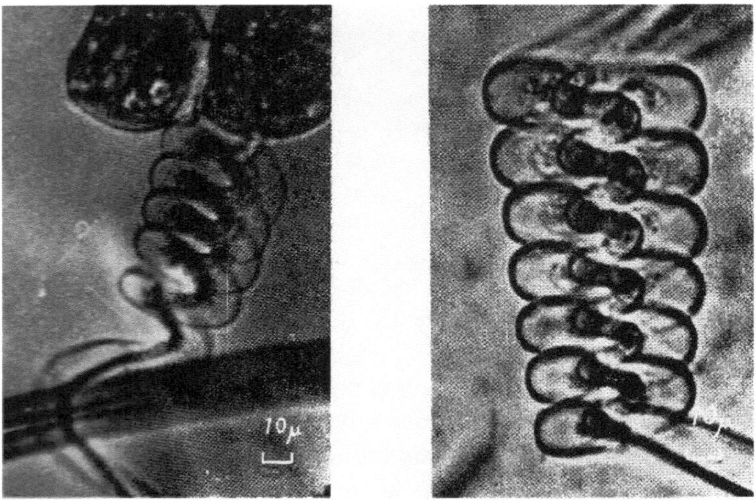

Figure 164 Photomicrographs of a protozoan animal *Carchesium* with its stalk in contracting state (Sugi, 1961).

My second good luck was my fortunate finding of local contractions spreading transversely around the perimeter of crayfish and frog muscle fibers. On this work, I would like to express my thanks to the late Professor Koji Uchizono of the University of Tokyo, who very generously let me work freely on my own idea, and to the late Professor Andrew F. Huxley for his interest in my work and encouragement. Almost certainly by his recommendation, my work was cited in the last edition of rofessor Hugh Davson's *A Textbook of General Physiology* (1970), together with my hand-drawn figure (Fig. 115).

From 1965 to 1967, I worked in Columbia University and the National Institutes of Health. I also wish to express my thanks to the late Professor Harry Grundfest, who generously allowed me to perform experiments on accommodation process of excitation–contraction coupling mechanism on my own idea, and to the late Dr. Richard J. Podolsky, who attracted my interest to the research field of muscle mechanics.

My third good luck was that in Teikyo University Medical School, where I was professor and chairman from 1973 to 2004, the circumstances around me were very favorable for receiving large scientific grants over the first 25 years, so that I could gather a number of brilliant people in my laboratory, whose names appear frequently in this volume. I list their names together with their final positions) here with sincere thanks:

- Dr. Teizo Tsuchiya (Professor, Kobe University)
- Dr. Takenori Yamada (Professor, Tokyo Science University)
- Dr. Takakazu Kobayashi (Professor, Shibaura Institute of Technology)
- Dr. Hidehiro Tanaka (Professor, Teikyo Junior College)
- Dr. Tsukasa Tameyasu (Associate Professor, St. Marianna University)
- Dr. Suechika Suzuki (Professor, Kanagawa Prefectural University)
- Dr. Hiroyuki Iwamoto (National Laboratory for Synchrotron Radiation Reaserarch)
- Dr. Kazukiro Oiwa (Laboratory Chief, National Laboratory for Scientific Research)
- Dr. Shigeru Chaen (Professor, Nihon University)
- Dr. Tsuyoshi Akimoto (President, Sanko IT, Inc.)
- Dr. Atsuko Suzuki (Professor, Yamanashi Health Science University)
- Dr. Seiryo Sugiura (Professor, The University of Tokyo)
- Dr. Yasutake Saeki (Professor, Tsurumi University).

Figure 165 An international symposium held at Hakone, Japan, in October 1986. Mount Fuji can be seen behind.

In addition to the wide-ranging research work done with my coworkers listed above, I could get grants to organize international symposia seven times in Tokyo, in Seattle, and mostly in Hakone, a most famous sightseeing spot in Japan (Figs. 165 and 166).

Figure 166 Professor Hugh E. Huxley and I during another international symposium held at Hakone, Japan, in October 2003.

The third, and the greatest, good luck struck me in the mid-1980s. As described in this volume, I had an idea to directly record myosin head movement, coupled with ATP hydrolysis, under a high-power electron microscope. At that time, Professor Akira Fukami had just succeeded in constructing the gas environmental chamber (EC) system after 10 years' painstaking efforts, and started looking for a coworker in the field of biology and physiology. As soon as I met him in his laboratory in Nihon University, we agreed to work together. As there are so many technical problems to be overcome, we made a number of trials and errors, until we published the first paper in *PNAS* in 1997 as communicated by Professor Andrew Huxley, who highly appreciated our EC experiments.

Here, I would like to express my hearty thank to President Kazuo Itoh of Japan Electron Optics Laboratory, Inc., for his interest in our work and generous financial support in the early period of our EC experiments (Fig. 167). In his young days, President Itoh constructed the first electron microscope in Japan together with the late Dr. Kenji Kazato, and they recently jointly received a prize in the USA.

My fantastic situation in Teikyo University was suddenly terminated in 1998 for a number of political reasons, and I lost my coworkers as well as my research money until the time of my retirement in 2004. Immediately after my retirement, however, I had the fourth good luck in keeping the space to resume EC experiments at the factory of Japan Electron Optics Laboratory

Figure 167 President Kazuo Itoh (left) and the electron microscope first constructed in Japan (right).

(JEOL) by kind and generous considerations of Professor Fukami and Dr. Terukazu Etoh, President of JEOL, one of Fukami's coworkers in developing the EC system (Fig. 168). Thus, I could perform EC experiments with Dr. Hiroyoshi Minoda in Tokyo Institute of Agriculture and Technology and his coworkers to record myosin head power and recovery strokes, while myosin filament samples were kindly provided by Professor Masaru Tanokura and Dr. Takuya Miyakawa in the University of Tokyo. I also thank them very much for their enthusiastic cowork, despite that the JEOL factory was geographically distant from their laboratories. Very sadly, Professor Fukami also passed away in 2006. It is my great regret that I could not show him the results of our EC experiments published in 2008, 2011, and 2015.

On the other hand, I had pleasant terms with Professor Andrew Huxley and his wife since we first met in 1965. Their last visit to my house was in 2010 (Fig. 169). Shortly after their visit, his wife Richenda died of cancer and Professor Huxley also passed away in 2012. He was always looking forward to having my experimental results on myosin head power stroke. Sadly, I cannot tell him my results anymore.

Meanwhile, my intimate relation with Professor Hugh Huxley was terminated by his sudden death in Woods Hole in

2013. In 2004, I invited Professor Huxley and his wife to my last international symposium held in Tokyo in March 2004, immediately after my retirement from Teikyo University (Fig. 170). It is very sad that about one month before his death, my wife and I were invited by him to his house in Concord, Massachusetts. It was also my great pity that I was unable to show him our results on myosin head power stroke.

Figure 168 President Terukazu Etoh of JEOL (left), Professor Akira Fukami (middle), and I (right). Mr. Hidehiro Ito (behind). Photograph taken at JEOL in 2004.

Figure 169 Professor A. F. Huxley and his wife (middle) during their last visit to my house in 2010. From left to right, Dr. Tsuyoshi Akimoto, My wife, Professor Andrew Huxley, Lady Huxley Richenda, Professor Rumiko Matsuoka (Tokyo Women's Medical College), and I.

My fifth good luck was the use of three different antibodies to myosin head; two of them were prepared by Drs. Kazuo Sutoh, Makihiro Tokunaga, and Takeyuki Wakabayashi, while the third one was prepared in my laboratory. These three antibodies enabled us to obtain evidence against the current dogma of muscle contraction mechanism as well as giving us a serious scepticism on in vitro motility assay systems in giving information about muscle contraction mechanism.

Figure 170 Photograph taken during music concert held during my last international symposium in Tokyo in 2004. From left to right, I, Professor Hugh Huxley, Mrs. Huxley, and my wife.

My sixth and probably the last good luck comes from the recent remarkable increase in the number of access-free scientific journals, obviously reflecting, at least in part, a number of flustrated investigators whose papers are unduly rejected by mainstream journals. In the past, I published my papers in a variety of mainstream journals. Recently, however, my papers related to the EC experiments are instantly rejected without reviewing procedures by such journals. I have already mentioned in this volume this terrible response of mainstream journals. I very much appreciate access-free journals for their fair manner of reviewing contributed papers. Due to the importance of access-free journals in promoting sound progress in science in general, and also the increasing benefits for publishers, access-free journals are now being published by even mainstream journal publishers.

Finally, I wish to thank Dr. Clive R. Bagshaw for his valuable comments on the early draft of this volume.

Haruo Sugi

References

Abbott BC, Aubert XM (1952) The force exerted by active striated muscle during and after change in length. *J Physiol*, 117: 77–86.

Amemiya Y, Iwamoto H, Kobayashi T, Sugi H, Tanaka H, Wakabayashi K (1988) Time-resolved X-ray diffraction studies on the effect of slow length changes on tetanized frog skeletal muscle. *J Physiol*, 407: 231–241.

Amemiya Y, Sugi H, Hashizume H (1979) X-ray diffraction studies on the dynamic properties of cross-bridges in Skeletal muscle. In *Cross-Bridge Mechanism in Muscle Contraction*, ed Pollack GH, Sui H, University of Tokyo Press, Tokyo, pp. 425–433.

Amemiya Y, Tameyasu T, Tanaka H, Hashizume H, Sugi H (1980) Time-resolved X-ray diffraction from frog skeletal muscle during shortening against an inertial load and a quick release. *Proc Jpn Acad*, 56B: 235–240.

Armstrong CM, Huxley AF, Julian FJ (1966) Oscillatory responses in frog skeletal muscle fibres. *J Physiol*, 186: 26P–27P.

Bagshaw CR (1993) *Muscle Contraction*, Chapman and Hall, London.

Bartels EM (2016) Muscle contraction revisited: Combining contraction models with present scientific research evidence. In *Muscle Contraction and Cell Motility: Fundamentals and Developments*, ed Sugi H, Pan Stanford Publishing, Singapore, pp. 75–116.

Bartels EM, Cooke PH, Elliott GF, Hughes RA (1993) The myosin molecule—chage response to nucleotide binding. *Biochim Biophys Acta*, 1157: 63–73.

Bartels EM, Elliott GF (1985) Donnan potentials from the A- and I-bands of glycerinated and chemically skinned muscles, relaxed and in rigor. *Biophys J*, 48: 61–76.

Buchthal F, Kaiser E (1951) The rheology of the cross-striated muscle fibre with special reference to isotonic conditions. *Dan Biol Med*, 21, nr.7.

Butler EP, Hale KF (1981) *Dynamic Experiments in the Electron Microscope. Practical Methods in Electron Microscopy*, vol 9, North-Holland, Amsterdam, pp. 457.

Chaen S, Shimada M, Sugi H (1986) Evidence for cooperative interaction of myosin heads with thin filament in the force generation of vertebrate skeletal muscle fibers. *J Biol Chem*, 261: 13632–13636.

Chaen S, Shirakawa I, Bagshaw CR, Sugi H (1997) Measurement of nucleotide release kinetics in single skeletal muscle myofibrils during isometric and isovelocity contractions using fluorescence microscopy. *Biophys J*, 73: 2033–2042.

Civan MM, Podolsky RJ (1966) Contraction kinetics of striated muscle fibres following quick changes in load. *J Physiol*, 184: 511–534.

Close RI (1972) The relation between sarcomere length and characteristics of isometric twitch contractions of frog sartorius muscle. *J Physiol*, 220: 745–762.

Cooke R (1981) Stress does not alter conformation of a domain of the myosin cross-bridge in rigor muscle fibres. *Nature*, 294: 570–571.

Cooke R, Crowder MS, Thomas DD (1982) Orientation of spin labels attached to cross-bridges in contracting muscle fibres. *Nature*, 300: 776–778.

Costantin LL (1970) The role of sodium current in the radial spread of contraction in frog muscle fibres. *J Gen Physiol*, 55: 703–715.

Ebashi S, Endo M (1968) Ca ion and muscle contraction. *Prog Biophys Mol Biol*, 18: 123–183.

Edman KAP (1979) The velocity of unloaded shortening and its relation to sarcomere length and isometric force in vertebrate muscle fibers. *J Physiol* 291: 143–159.

Edman KAP (1988) Double hyperbolic force-velocity relation in frog muscle fibres. *J Physiol*, 404: 301–321.

Edman KAP, Curtin NA (2001) Synchronous oscillations of length and stiffness during loaded shortening of frog muscle fibres. *J Physiol*, 534: 553–563.

Eisenberg E, Moos C (1968) The adenosine triphosphatase activity of acto-heavy meromyosin. A kinetic analysis of actin activation. *Biochemistry*, 7: 1486–1489.

Eisenberg E, Moos C (1970) Actin activation of heavy meromyosin adenosine triphosphatase, Dependence on adenosine triphosphate and actin concentrations. *J Biol Chem*, 245: 2451–2456.

Elliott GF, Lowy J, Worshington CR (1963) An X-ray and light diffraction study of the filament lattice of striated muscle in the living state and in rigor. *J Mol Biol*, 6: 295–305.

Fenn WO (1923) A quantitative comparison between the energy liberated and the work performed by the isolated sartorius of the frog. *J Physiol*, 58: 441–454.

Finer JT, Simmons RM, Spudich JA (1994) Single myosin molecule mechanics: Piconewton forces and nanometre steps. *Nature*, 368: 113–119.

Fisher AJ, Smith CA, Thoden JB, Smith R, Sutoh K (1995) X-ray structures of the myosin motor domain of *Dictyostelium discoideum* complexed with MgADP.BeFx and MgADP.AlF4. *Biochemistry*, 34: 8960–8972.

Ford LE, Huxley AF, Simmons RM (1981) The relation between stiffness and filament overlap in stimulated frog muscle fibres. *J Physiol*, 311: 219–249.

Fukami A, Adachi K (1965) A new method of preparation of a self-perforated microplastic grids and its applications. *J Electron Microsc*, 14: 112–118.

Fukushima K (1987) Production and application of the gas environmental chamber (in Japanese), doctoral thesis, Nihon University, Tokyo.

Fulton JF (1925) On the summation of contractions in skeletal muscle. *Am J Physiol*, 75: 211–234.

Geeves MA, Holmes KC (2005) The molecular mechanism of muscle contraction. *Adv Protein Chem*, 71: 161–193.

Geeves MA, Lehrer SS (1994) Dynamics of the muscle thin filament regulatory switch: The size of the cooperative unit. *Biophys J*, 67: 273–282.

Gordon AM, Godt RE, Donaldson KB, Harris CE (1973) Tension in skinned frog muscle fibers in solutions of varying ionic strength neutral salt concentration. *J Gen Physiol*, 62: 550–574.

Hanson J, Huxley HE (1953) Structural basis of the cross-striations in muscle. *Nature*, 172: 530–532.

Hartree W, Hill AV (1921) The nature of the isometric twitch. *J Physiol*, 55: 389–411.

Haselgrove JC, Huxley HE (1973) X-ray evidence for radial cross-bridge movement and for the sliding filament model in actively contracting skeletal muscle. *J Mol Biol*, 77: 549–568.

Hatta I, Sugi H, Tamura Y (1988) Stiffness changes in frog skeletal muscle during contraction recorded using ultrasonic waves. *J Physiol*, 403: 193–209.

He ZH, Chillingsworth RK, Brune M, Corrie JET, Webb MR, Ferenczi MA (1997) The efficiency of contraction in rabbit skeletal muscle fibers determined from the rate of release of inorganic phosphate. *J Physiol*, 517: 839–854.

Hill AV (1938) The heat of shortening and the dynamic constants of muscle. *Proc Roy Soc B*, 126: 136–195.

Hill, AV (1939) The mechanical efficiency of frog's muscle. *Proc Roy Soc B*, 127: 434–451.

Hill AV (1964a) The variation of total heat production in a twitch with velocity of shortening. *Proc Roy Soc B*, 159: 596–605.

Hill AV (1964b) The efficiency of mechanical power development during muscular shortening and its relation to load. *Proc Roy Soc B*, 159: 319–324.

Hill AV, Howarth JV (1959) The reversal of chemical reactions in contracting muscle during an applied stretch. *Proc R Soc B*, 151: 169–193.

Hodgkin AL, Horowicz P (1960) Potassium contractures in single muscle fibres. *J Physiol*, 153: 386–403.

Holmes KC (1995) The actomyosin interaction and its control by tropomyosin. *Biophys J*, 68: 2s–7s.

Huxley AF (1957) Muscle structure and theories of contraction. *Prog Biophys Biophys Chem*, 7: 255–318.

Huxley AF (1998) Support for the lever arm. *Nature*, 396: 317–318.

Huxley AF, Niedergerke R (1954) Interference microscopy of living muscle fibres. *Nature*, 173: 971–973.

Huxley AF, Peachey LD (1961) The maximum length for contraction in vertebrate striatedmuscle. *J Physiol*, 156: 150–165.

Huxley AF, Simmons RM (1971) Proposed mechanism of force generation in striated muscle. *Nature*, 233: 533–538.

Huxley AF, Simmons RM (1972) Mechanical transients and the origin of muscular force. *Cold Spring Harb Symp Quant Biol*, 27: 669–680.

Huxley AF, Taylor RE (1958) Local activation of striated muscle fibres. *J Physiol*, 144: 426–441.

Huxley HE (1953a) X-ray analysis and the problem of muscle. *Proc Roy Soc B*,141: 59–66.

Huxley HE (1953b) Electron microscope studies of the organization of the filaments in striated muscle. *Biochim Biophys Acta*, 12: 387–394.

Huxley HE (1957) The double array of filaments in cross-striated muscle. *J Biochem Biophys Cytol*, 3: 631–648.

Huxley HE (1968) Structural difference between resting and rigor muscle. *Proc Roy Soc B*, 141: 59–62.

Huxley HE (1969) The mechanism of muscular contraction. *Science*, 164: 1356–1366.

Huxley HE (1990) Sliding filaments and molecular motile systems. *J Biol Chem*, 265: 8347–8350.

Huxley HE, Brown W (1967) The low angle X-ray diagram of vertebrate striated muscle and its behaviour during contraction and rigor. *J Mol Biol*, 30: 384–434.

Huxley HE, Hanson J (1954) Changes in the cross-striations of muscle during contraction and stretch and their structural interpretation. *Nature*, 173: 973–976.

Huxley HE, Simmons RM, Faruqi AR, Kress M, Bordas J, Koch MHJ (1983) Changes in the X-ray diffractions from contracting muscle during rapid mechanical transients and their structural implications. *J Mol Biol*, 169: 469–506.

Huxley HE, Stewart A, Sosa H, Irving T (1994) X-ray diffraction measurements of the extensibility of actin and myosin filamentsin contracting muscle. *Biophys J*, 67: 2411–2421.

Ishijima A, Doi T, Sakurada K, Yanagida T (1991) Sub piconewton force fluctuations of actomyosin in vitro. *Nature*, 352: 301–306.

Iwamoto H, Sugaya R, Sugi H (1990) Force-velocity relation of frog skeletal muscle fibres shortening under continuously changing load. *J Physiol*, 422: 185–202.

Julian FJ, Morgan DL (1979) Intersarcomere dynamics during fixed-end tetanic contractions of frog muscle fibres. *J Physiol*, 293: 365–378.

Julian FJ, Sollins MR (1975) Variation of muscle stiffness with force at increasing speeds of shortening. *J Gen Physiol*, 66: 287–302.

Kaya M, Higuchi H (2010) Nonlinear elasticity and an 8 nm working stroke of single myosin molecules in myofilaments. *Science*, 329: 686–689.

Kensler RW, Stewart M (1983) Frog skeletal muscle thick filaments are three- stranded. *J Cell Biol*, 96: 1797–1802.

Kraft TM, Messecrli M, Rsen-Rutschauer B, Perriad JC, William T, Brenner B (1995) Equilibrium and exchange of fluorescently labeled molecules in skinned skeletal muscle fibers visualized by confocal microscopy. *Biophys J*, 69: 1246–1258.

Lehrer (1994) The regulatory switch of the muscle thin filament: Ca^{2+} or myosin heads? *J Muscle Res Cell Motil*, 15: 232–236.

Levine BA, Moir AJR, Trayer LP, Williams RJP (1990) In *Molecular Mechanism of Muscle Contraction*, ed Squire JM., Macmillan Press, London, pp. 171–209.

Lymn RW, Taylor EW (1971) Mechanism of adenosine triphosphate hydrolysis by actomyosin. *Biochemistry*, 16: 4617–4624.

Mau MC, Rowe AJ (1980) Fraying of A-filaments into three subfilaments. *Nature*, 286: 412–414.

Milligan RA, Whittaker M, Safer D (1990) Molecular structure of F-actin and location of surface binding sites. *Nature*, 348: 217–221.

Minoda H, Okabe T, Inayoshi Y, Miyakawa T, Miyauchi Y, Tanokura M, Katayama E, Wakabayashi T, Akimoto T, Sugi H (2011) Electron microscopic evidence for the myosin head lever arm mechanism in hydrated myosin filaments using the gas environmental chamber. *Biochem Biophys Res Commun*, 405: 651–656.

Muhlrad A, Peyser YM, Nili M, Ajtai K, Reisler E, Burghardt TP (2003) Chemical decoupling of ATPase activation and force production from the contractile cycle in myosin by steric hindrance of lever arm movement. *Biophys J*, 84: 1047–1056.

Muretta JM, Rohde JA, Johnsrud DO, Cornea S, Thomas DD (2015) Direct real-time detection of the structural and biochemical events in the myosin power stroke. *Proc Natl Acad Sci U S A*, 112: 14272–14277.

Nakajima S, Bastian J (1976) Membrane properties of the transverse tubular system of amphibian skeletal muscle. In *Electrophysiology of Nerve Synapse and Muscle*, ed Reuben JP, Purpura DP, Bennett MVL, Kandel ER, Raven Press, New York, pp. 242–267.

Naylor GRS, Podolsky RJ (1981) X-ray diffraction of striated muscle fibers in rigor. *Proc Natl Acad Sci U S A*, 78: 5559–5563.

Nishiyama K (1984) Analysis of mechanical behaviour of muscle by a multi-sarcomere model. In *Contractile Mechanisms in Muscle*, ed Pollack GH, Sugi H, Plenum, New York, pp. 637–639.

Nishizaka T, Seo R, Tadakuma H, Kinosita K, Ishiwata S (2000) Characterization of single actomyosin rigor bonds: Load dependence of lifetime and mechanical properties. *Biophys J*, 79: 962–974.

Oiwa K, Chaen S, Kamitsubo E, Shimmen T, Sugi H (1990) Steady-state force-velocity relation in the ATP-dependent sliding movement of myosin-coated beads on actin cables in vitro studied using a centrifuge microscope. *Proc Natl Acad Sci U S A*, 87: 7893–7897.

Oiwa K, Chaen S, Sugi H (1991) Measurement of work done by ATP-induced sliding between rabbit muscle myosin and algal cell actin cables in vitro. *J Physiol*, 437: 751–763.

Oiwa K, Kawakami T, Sugi H (1993) Unitary distance of actin-myosin sliding studied using an in vitro force-movement assay system. *J Biochem*, 114: 28–32.

Oplatka A (1997) Critical review of the swinging crossbridge theory and of the cardinal active role of water in muscle contraction. *Crit Rev Biochem Mol Biol*, 32: 307–360.

Peachey LD (1965) The sarcoplasmic reticulum and transverse tubules of frog's sartorius. *J Cell Biol*, 25: 209–231.

Radocaj A, Weiss T, Helsby WI, Brenner B, Kraft T (2009) Force-generating cross-bridges during ramp-shaped releases: Evidence for a new structural state. *Biophys J*, 96: 1430–1446.

Ramsey RW, Street SF (1941) Muscle function as studied in single muscle fibres. *Biol Symp*, 3: 9–34.

Rayment I, Holden HM, Whittacker M, Yohn CB, Lorenz M, Holmes KC, Milligan RA (1993a) Structure of the actin-myosin complex and its implications for muscle contraction. *Science*, 261: 58–65.

Rayment I, Rypniewski WR, Schmidt-Base K, Smith R, Tomchick DR, Benning MM, Winkelman DA, Wesenberg G, Holden HM (1993b) Three-dimensional structure of myosin subfragment-1: A molecular motor. *Science*, 261: 50–58.

Reggiani C, Potma FJ, Bottinelli R, Canepari M, Pellegrino MA, Stienen GJM (1997) Chemo-mechanical energy transduction in relation to myosin isoform composition in skeletal muscle fibers of the rat. *J Physiol*, 502: 449–460.

Reisler S (1980) On the question of co-operative interaction of myosin heads with F-actin in the presence of ATP. *J Mol Biol*, 138: 93–107.

Schoenberg M, Wells JB, Podolsky RJ (1974) Muscle compliance and the longitudinal transmission of mechanical impulse. *J Gen Physiol*, 64: 623–642.

Sheez MP, Spudich MJ (1983) Movement of myosin-coated fluorescent beads on actin cables in vitro. *Nature*, 303: 31–35.

Shimmen T, Yano M (1984) Active sliding movement of latex beads coated with skeletal muscle myosin on *Chara* actin cables. *Protoplasma*, 121: 132–137.

Squire JM (1972) General model of myosin filament structure II. Myosin filaments and cross-bridge interactions in vertebrate striated and insect flight muscles. *J Mol Biol*, 72: 125–138.

Suda H, Ishikawa A, Fukami A (1992) Evaluation of the critical electron dose on the contractile activity of hydrated muscle fibers in the film-sealed environmental cell. *J Electron Microsc*, 41: 33–39.

Sugi H (1960) Propagation of contraction in the stalk muscle of *Carchesium*. *J Fac Sci Univ Tokyo IV*, 8: 603–615.

Sugi H (1992) *Muscle Contraction and Cell Motility, Molecular and Cellular Aspects, Advances in Comparative and Environmental Physiology*, vol 12, Springer, Berlin.

Sugi H (1993) Molecular mechanism of ATP-dependent actin-myosin interaction in muscle contraction. *Jpn J Physiol*, 43: 435–454.

Sugi H (1998) *Current Methods in Muscle Physiology, Advantages, Problems and Limitations*, Oxford University Press, Oxford.

Sugi H (2016) *Muscle Contraction and Cell Motility, Fundamentals and Developments*, Pan Stanford Publishing, Singapore.

Sugi H, Abe T, Kobayashi T, Chaen S, Ohnuki Y, Saeki Y, Sugiura S (2013) Enhancement of force generated by individual myosin heads in skinned rabbit psoas muscle fibers at low ionic strength. *PLoS ONE*, 8: e63658.

Sugi H, Akimoto T, Sutoh K, Chaen S, Oishi N, Suzuki S (1997) Dynamic electron microscopy of ATP-induced myosin head movement in living muscle thick filaments. *Proc Natl Acad Sci U S A*, 94: 4378–4382.

Sugi H, Chaen S, Akimoto T, Minoda H, Miyakawa T, Miyauchi Y, Tanokura M, Sugiura S (2015) Electron microscopic recording of myosin head power stroke in hydrated myosin flaments. *Sci Rep*, 5: 15700.

Sugi H, Chaen S, Kobayashi T, Abe T, Kimura K, Saeki Y, Ohnuki Y, Miyakawa T, Tanokura M, Sugiura S (2014) Definite differences between in vitro actin-myosin sliding and muscle contraction as revealed using antibodies to myosin head. *PLoS ONE*, 9: e93272.

Sugi H, Iwamoto H, Akimoto T, Kishi H (2003) High mechanical efficiency of the cross-bridge powerstroke in skeletal muscle. *J Exp Biol*, 206: 1201–1206.

Sugi H, Iwaoto H, Akimoto T, Ushitani H (1998) Evidence for the load-dependent mechanical efficiency of individual myosin heads in skeletal muscle fibers activated by laser flash photolysis of caged calcium in the presence of a limited amount of ATP. *Proc Natl Acad Sci U S A*, 95: 2273–2278.

Sugi H, Kobayashi T (1983) Sarcomere length and tension changes in tetanized frog frog muscle fibers after quick stretches and releases. *Proc Natl Acad Sci U S A*, 80: 6422–6425.

Sugi H, Kobayashi T (2006) Remaining mysteries about the sliding filament mechanism in muscle contraction. *Recent Res Devel Biophys*, 5: 167–186.

Sugi H, Kobayashi T, Gross T, Noguchi K, Karr T, Harrington WF (1992) Contraction characteristics and ATPase activity of skeletal muscle fibers in the presence of antibody to myosin subfragment-2. *Proc Natl Acad Sci U S A*, 89: 6134–6137.

Sugi H, Kosaka K (1964) Summation of contraction in single crayfish muscle fibres. *Jpn J Physiol*, 14: 450–467.

Sugi H, Minoda H, Inayoshi Y, Yumoto F, Miyakawa T, Miyauchi Y, Tanokura M, Akimoto T, Kobayashi T, Chaen S, Sugiura S (2008) Direct demonstration of the cross-bridge recovery stroke in muscle thick filaments in aqueous solution by using the hydration chamber. *Proc Natl Acad Sci U S A*, 105: 17396–17401.

Sugi H, Ochi R (1967a) The mode of transverse spread of contraction initiated by local activation in single crayfish muscle fibers. *J Gen Physiol*, 50: 2145–2166.

Sugi H, Ochi R (1967b) The mode of transverse spread of contraction initiated by local activation in single frog muscle fibers. *J Gen Physiol*, 50: 2167–2176.

Sugi H, Tsuchiya T (1981a) Isotonic velocity transients in frog muscle fibres following quick changes in load. *J Physiol*, 319: 239–252.

Sugi H, Tsuchiya T (1981b) Enhancement of mechanical performance in frog muscle fibres after quick increases in load. *J Physiol*, 319: 239–252.

Sugi H, Tsuchiya T (1988) Stiffness changes during enhancement and deficit of isometric force by slow length changes in frog skeletal muscle fibres. *J Physiol*, 407: 215–229.

Sugi H, Tsuchiya T (1998) Muscle mechanics I, Intact single muscle fibres. In *Current Methods in Muscle Physiology, Advantages, Problems and Limitations*, ed Sugi H, Oxford University Press, Oxford, pp. 3–31.

Sugi H, Yamaguchi M, Ohno T, Kobayashi T, Chaen S, Okuyama H (2016) Tension recovery following ramp-shaped release in high-Ca and low-Ca rigor muscle fibers: Evidence for the dynamic state of AMADP myosin heads in the absence of ATP. *PLos ONE*, DOI: 10.1371/journal pone 0162003.

Sugiura S, Kobayakawa N, Fujita H, Yamashita H, Momomura S, Chaen S, Omata M, Sugi H (1998) Comparison of unitary displacements and forces between 2 cardiac myosin isoforms by the optical trap technique: Molecular basis for cardiac adaptation. *Circ Res,* 82: 1029–1034.

Sun YB, Hilber K, Irving M (2001) Effect of active shortning on the rate of ATP utilization by rabbit psoas muscle fibers. *J Physiol,* 531: 781–791.

Sutoh K, Tokunaga M, Wakabayashi T (1989) Electron microscopic mapping of myosin head with site-directed antibodies. *J Mol Biol,* 206: 357–363.

Suzuki S, Oshimi Y, Sugi H (1993) Freeze-fracture studies on the cross-bridge angle distribution at various states and the thin filament stiffness in single skinned frog muscle fibers. *J Electron Microsc,* 42: 107–116.

Suzuki S, Pollack GH (1986) Bridgelike interconnections between thick filaments in stretched skeletal muscle fibers observed by the freeze-fracture method. *J Cell Biol,* 102: 1093–1098.

Suzuki S, Sugi H (1983) Extensibility of the myofilaments in vertebrate skeletal muscle as revealed by stretching rigor muscle fibers. *J Gen Physol,* 81: 531–546.

Sweeny HL, Houdusse A (2010) Structural and functional insights into the myosin motor mechanismm. *Ann Rev Biophys,* 39: 539–557.

Szent-Györgyi A. (1951) *Chemistry of Muscular Contraction.* Academic Press, New York, p. 162.

Takahashi I, Oiwa K, Kawakami T, Tanaka H, Sugi H (1993) Scanning electron microscopy of the myosin-coated surface of polystyrene beads in a force-movement assay system for ATP-dependent actin-myosin sliding. *J Electron Microsc,* 42: 334–337.

Tamura Y, Hatta I, Matsuda T, Sugi H, Tsuchiya T (1982) Changes in muscle stiffness during contraction recorded using ultrasonic waves. *Nature,* 299: 631–633.

Tanaka H, Sugi H, Hashizume H (1984) Effect of stretch on the equatorial X-ray diffraction pattern from frog skeletal muscle I rigor. In *Contractile Mechanisms in Muscle,* ed Pollack GH, Sugi H, Plenum New York, pp. 203–205.

Thames MD, Teichholz LE, Podolsky RJ (1974) Ionic strength and the contraction kinetics of skinned muscle fibers. *J Gen Physiol,* 63: 509–530.

Tokiwa T, Morales MF (1971) Independent and cooperative reactions of myosin heads with F-actin in the presence of adenosine triphosphate. *Biochemistry*, 10; 1722–1727.

Tonomura Y (1972) *Muscle Protein, Muscle Contraction and Cation Transport*. University of Tokyo Press, Tokyo.

Trybus KM, Taylor EW (1980) Kinetics of the cooperative binding of subfragment 1 to regulated actin. *Proc Natl Acad Sci U S A*, 77: 7209–7213.

Tsuchiya T, Tanaka H, Shirakawa I, Karr T, Sugi H (1998) Evidence for the essential role of myosin subfragent-2 in the ATP-dependent actin-myosin sliding in muscle contraction. *Jpn J Physiol*, 48: 383–387.

White DCS, Thorson J (1973) The kinetics of muscle contraction. *Prog Biophys Mol Biol*, 27: 175–255.

Wolege RC, Curtin NA, Homsher E (1985) *Energetic Aspects of Muscle Contraction*. Academic Press, London.

Yanagida T (1984) Angles of fluorescently labelled myosin heads and actin monomers in contracting and rigor stained muscle fiber. In *Contractile Mechanisms in Muscle*, ed Pollack GH, Sugi H, Plenum, New york, pp. 397–411.

Yanagida T (1985) Angle of active site of myosin heads in contracting muscle during sudden length changes. *J Musce Res Cell Motil*, 6: 43–52.

Index

abdominal extensor muscle 203
actin 2–7, 14–17, 20, 28–35, 50–52, 54, 57, 59, 62, 65, 109–110, 113–115, 124, 126, 187
  arrangement of 3–4
  mixture of 3, 113–114
  physiological function of 65–66
actin-binding sites 28, 30, 32, 86, 95, 98, 101, 104, 132
actin cables 34–38, 102–104, 149, 152–153
actin filament, fluorescent 33
actin filament arrays 38
  well-organized 34
actin filament axis 124
actin filament constant 46
actin filament displacement, stretch-induced 180
actin filament elasticity 19
actin filament extensibility 17–18
actin filament lattice 180
actin filament length 17
actin filament sliding 33, 87, 90
  zigzag 88
actin filament stiffness 17
actin filaments 5–8, 15, 31–32, 38–40, 45–46, 50–54, 56–59, 87–88, 97–98, 104–106, 113–115, 122–133, 135–137, 174–175, 180
actin monomers 7, 18–19, 128
  axial spacing of 19
actin–myosin binding, strong 101
actin–myosin interaction 31, 135–136, 149
  cyclic 135
  electrostatic 135
  kinetics of 144–145
actin–myosin linkages 134
  electrostatic 135
  rigor 118
actin–myosin rigor complex 86
actin–myosin sliding 85, 88, 97–98, 102
actin–myosin sliding velocity 102
action potential, generation of 224
actomyosin 2, 28
ADP 28–30, 78, 80, 86–87, 126, 129–131, 155–157
A–M links 10–11
anomalous local contraction 223
anti-LD antibody 66, 90–92
antibodies 44, 66–68, 82–83, 85–111, 114, 116–119, 122–125, 132, 247
  anti-CAD 66, 87
  anti-COD 66
  anti-RLR 89
  anti-S-2 98–102, 104–105, 110, 137
  anti-S2 109
  bulky 86
  inhibitory effect of 91, 97–98

polyclonal 99
site-directed monoclonal 66
ATP 28–30, 62–66, 72, 75, 77–78, 80, 83, 121, 124, 126–132, 135–136, 149, 152–153, 155–156, 169–170
absence of 16, 31, 34, 70–71, 104, 127, 129–130
exhaustion of 80–81, 129
ATP-activated myosin heads 117–118, 123
ATP-dependent actin filament sliding 88
ATP-dependent actin–myosin sliding 88
ATP-dependent bead movement 37–38
ATP-dependent movement of fluorescent actin filaments 33
ATP-dependent sliding of actin filaments 87
ATP-dependent sliding velocity of myosin-coated beads 105
ATP hydrolysis 3, 28, 58, 68, 80, 93, 97, 99, 113, 126, 129, 135, 140, 170, 173
chemical energy of 129, 131
ATP-induced actin–myosin sliding 35
ATP-induced myosin head movement 30, 62, 72, 74, 76, 78, 80, 82, 118, 121–122, 130
electron microscopic recording of 61–62, 64, 66, 68, 70, 72, 74, 76, 78, 80, 82, 84
ATP-induced myosin head power and recovery strokes 125, 127, 129, 131
ATP-induced myosin head power stroke 113–115, 117–118, 123, 125, 127, 130
reversibility of 118
ATP-induced myosin head recovery stroke 75, 129
ATP-induced sliding 34–35
ATP scavenger 34, 80, 118, 152–153
ATPase activity 28, 33, 88–89, 93, 97, 149
actin-activated 88, 98
attachment–detachment cycle 7–8, 15, 30, 40, 109, 119, 126, 132
auxotonic load 158–162, 164–165
moderate 161–162
auxotonic shortening 158–159, 161

bead movement 36–37, 102
binding, myosin head–actin 136

CAD, *see* catalytic domain
catalytic domain (CAD) 43–44, 46, 82, 97–98, 123, 125
CC, *see* contractile component
COD, *see* converter domain
contractile activity 93, 140–141
contractile component (CC) 8–9, 190–191, 193
contractile mechanisms 191, 220
contractile proteins 2–3, 5, 191
contractile system 133
contracting muscle, stiffness 134

contraction
  graded 195, 212, 239
  localized 191
  perceptible 214, 221
  phasic 238
  rapid phasic 238
converter domain (COD) 43–47, 83, 89, 97–98, 124–125
Cy3-EDA-ATP 149–150
Cy3-EDA-nucleotide 149, 151
  displacement of 149–151

deep-etch replica method 49–50, 52, 54, 56, 58
depolarization 191, 205, 207, 211, 221, 225, 228, 230–231, 236–238
depolarized fiber surface 227–228, 230, 233–237
disordering, stretch-induced 180, 183
dogfish jaw muscle, contracting 172

EC, *see* environmental chamber
EC experiments 68, 71–72, 74, 76, 80–81, 84–85, 111, 113, 121, 123–125, 128, 130–131, 136, 244–245, 247
electron micrographs 4, 53, 68, 110, 114, 182–183
electron microscope 4, 31, 62–63, 245
electrostatic repulsion forces 180–181
environmental chamber (EC) 57–58, 62–66, 72, 82, 84–85, 132, 244
enzyme kinetics 28
excitation–contraction coupling 208, 219–221, 224, 242

F-actin 33
Fenn Effect 139–140, 142, 144, 146, 148–153
fiber
  deteriorated 196
  flash-activated 158
  low-rigor 95
  nerve 143, 218
  p-PDM-inactivated 110–111
  p-PDM-treated 108–109
  tetanized 174, 176, 182, 185
fiber shortening 23, 57, 109, 148, 158–159, 161, 164–165
  auxotonic 160, 163
  early phase of 160, 164
  flash-induced 159, 164
  flash-induced isotonic 166
fiber slack length 176, 179
filament backbone 6, 10, 15
filament sliding 7, 10
fluorophores 61
force–velocity equation 141
force–velocity relations 86, 91–92, 120, 143–145, 147, 174, 190
Fourier transform diffractograms 110–111, 182
freeze-etch replica images 50–51
frog muscle fibers 193, 212–215, 217, 219, 242
fused isometric response 198

gas environmental chamber 61–63, 65, 67
glass microneedles, myosin-coated 34–35, 152
gold particles 67–71, 74–81, 116, 118–119, 122
graded local contraction 214, 226, 235–237

heavy meromyosin (HMM) 5, 28, 33, 38–41, 88, 97
high-Ca rigor fibers 95–97
Hill equation 143
HMM, *see* heavy meromyosin
Huxley, A. 47, 218, 242, 244–246
Huxley, H. 82, 245, 247
Huxley's contraction model 8–11, 20, 57, 80, 115, 139
Huxley–Simmons contraction model 13, 15–16, 20, 33, 43
Huxley–simmons model 13–14, 16, 18, 20, 22, 24

I-band width 5
I-bands 2, 17, 19, 24, 57, 136, 193
isometric contraction 149–150, 169, 195
isometric responses 196, 198
isometric tetanus 173, 177, 182–184, 187
isometric twitch 190–191, 193–194
isotonic shortening 156, 165, 167–170, 238
  afterloaded 165, 167
isotonic velocity transients 174–176

local contraction 193, 205–213, 216–218, 220, 224–228, 230–238, 242
  reversible 205
local membrane depolarization 203, 205, 211–212, 218–221, 223–224, 226, 235
longitudinal muscle stiffness 134
Lymn–Taylor scheme 27, 29–30, 33, 78, 86, 157, 159
lysine residue 98
  chemical modification of 98
  reactive 66, 82, 88–89, 97–98

M-ADP-Pi 28, 30, 78, 80, 126, 129–131, 136, 157, 161, 164–165, 169
M-ATP 30, 78, 129–130
mechanical efficiency of individual myosin heads 155, 157, 165, 170
  measurement of 157, 159, 161, 163, 165, 167, 169
mechanical efficiency of individual myosin heads during auxotonic shortening 157
mechanical efficiency of individual myosin heads during isotonic shortening 165
mechanical efficiency of myosin heads on isotonic load 169

mechanical impulse 21, 23–24, 187
membrane depolarization 192, 195, 206, 208, 211–212, 219, 226, 228, 235, 237–238
  local 205, 214, 226–227
  moderate 205–206, 211–213, 225
membrane resistance 208, 211, 213, 220
  reduction of 211, 213, 219
meridional reflections 110–111
MgATPase activity 93, 98–99, 106–108, 120
molluscan paramyosin filaments 68
motility assay systems 33, 35–38, 41, 72, 97, 148–149, 151–153, 247
muscle
  intact 20, 181
  relaxed 30, 113, 126–128, 157
  resting 1
  semitendinosus 212, 225
  stimulated 191
  stretched 178, 182
  tetanized 173, 186
muscle biochemistry 33
muscle biochemists 61
muscle contraction 1, 4–5, 7–8, 24, 27–28, 30, 43, 85, 87–88, 94–95, 97–99, 103–107, 135–136, 155–156, 189–193
  effect of antibodies on 94–95, 97, 99, 101, 103, 105, 107, 109, 111
  molecular mechanism of 7, 27, 49, 135
  role of regenerative process in 238
muscle crystallographists 74, 84
muscle fiber contraction 85, 88
muscle fiber elasticity 15–16
muscle fiber membrane 238
muscle fiber shortening 145, 178
muscle fiber stiffness 101, 108–109, 176–178, 181, 184–185
muscle fiber stiffness in single tetanized muscle fibers, measurement of 176–177, 184–185
muscle fibers
  activated 86, 90–93, 100, 109, 120
  crayfish 195, 197–198, 208, 212
  frog 203, 219
  frozen 54
  intact 16, 186
  intact contracting 95
  intact frog 145
  K-contractures of 219
  low-Ca rigor 95
  mammalian 193
  p-PDM-inactivated 110
  p-PDM-treated 106–109
  single 14, 34, 143–145, 147, 155, 192, 203, 212–213
  single skeletal 14
  single skinned 18, 96, 149, 155, 158, 165–167
  single tetanized 177, 184
  skinned frog 50
  skinned skeletal 86, 106
  suspended 196
  suspended crayfish 196
  tetanized 175–176, 184, 186, 235, 239
muscle longitudinal stiffness 134
muscle mechanics 24, 191, 242
muscle myofibrils 149

muscle protein 144
muscle shortening 137, 140, 142, 145, 174
  steady 11
muscle slack length 9
muscle stiffness 132–135, 178
  measurement of 133
muscle transverse stiffness 134
muscle twitch 190
myofibrils 2, 5, 17, 65–66, 148–151, 153, 193
myofilament elasticity 17, 20
myofilament extensibility 16–17
myofilament lattice 27–28, 32–33, 45, 49–50, 52, 61, 88, 97, 110, 127, 136, 149, 180–183
myofilament-lattice 97, 159
myofilament lattice
  constant-volume 181
  hexagonal 3, 180, 183
  stretch-induced disorder of 182–183
myofilament sliding 21, 47, 82, 115, 123
myofilaments 16–17, 19–20, 49–50, 54, 65, 110, 135, 144
  living muscle 58
myosin 2–3, 5–6, 15–16, 28–29, 32–35, 50–51, 53, 56, 58, 66–67, 87–88, 90, 102–105, 107, 136
  attachment of 38
  full-length 40, 97
myosin-coated beads 36, 102, 105
myosin filament axis 54, 76, 113, 127
myosin filament backbone 6, 15, 44–45, 47, 97, 125
myosin filament extension 17
myosin filaments 3–8, 15–17, 37–38, 46–47, 50–54, 56–57, 65–66, 74–75, 78–79, 110–111, 113–117, 126–127, 129–131, 183, 187–188
  packed 180
  spindle-shaped 114
  synthetic 64, 74, 80, 94, 114, 116
  vertebrate muscle 69, 72, 74
myosin head-actin binding strength, regulation of 102
myosin head-actin molecule complex 32
myosin head CAD 45–46, 66, 76, 82–83, 86, 88, 93, 95, 97–99, 102, 104, 109–110, 123–124
  attachment angle of 124
  distal region of 82, 117, 123–124
  proximal region of 82, 117, 122–125
myosin head CAD regions 101–102
myosin head COD 66, 82, 88–89, 97–98
myosin head concentration 157–158, 169
myosin head configuration 46–47, 58–59, 125
myosin head crystals 27, 44–45
myosin head LD 45–47, 61, 66, 82, 90–93, 98–99, 124, 165, 167
myosin head movement 8, 58, 62, 68, 72, 76–77, 80–83, 127, 129, 244
myosin head orientation 21, 98
myosin head performance 38, 171

myosin head polarity 74, 79
myosin head position 71, 73, 75–78, 80, 126, 128, 131
myosin head power stroke 14, 16, 33, 44, 46–47, 85, 113–114, 116–118, 120, 122–128, 130, 132, 134, 136, 245–246
  amplitude of 16, 33, 74, 123–124, 167
  electron microscopic recording of 113–114, 116, 118, 120, 122, 124, 126, 128, 130, 132, 134, 136
  mode of 125
  modes of 124
myosin head recovery stroke 81–82, 115, 122, 124
myosin head structure 44, 58, 98, 125
myosin head tilting 13, 20–21, 43, 57, 59
myosin head–actin filament binding 136
myosin head–actin filament interface 136
myosin heads 6–8, 15–16, 30–35, 37–40, 50–54, 56–59, 68–72, 77–80, 82–83, 85–88, 90–98, 106–111, 113–119, 123–132, 157–159
  actin-activated 88–89, 97
  enhanced performance of 119, 121, 123
  fixed 34, 88–89
  individual 40, 65–66, 76–78, 114–116, 118, 120–123, 125, 153, 155, 157, 159, 161, 163, 165, 169–170
  load-dependent mechanical efficiency of 155–158, 160–162, 164, 166, 168, 170
  pairs of 50, 52
  redistribution of 14, 175
  single 38–39
myosin molecules 5–6, 27, 33–34, 72, 87–88, 97, 102, 104–110
myosin rod 74, 136
myosinmyosin rod cofilament 74
myosin–paramyosin core 68, 74
myosin–paramyosin core complex 68–69, 71, 73

NADH 99, 155–156
NMR, *see* nuclear magnetic resonance
nuclear magnetic resonance (NMR) 155

optical trap method 38, 40–41, 98
optical trap system 38–41

p-PDM 106–107, 109, 111
p-PDM modification 110
paramyosin filaments, thick 68
piezoelectric transducer 133
polystyrene beads, myosin-coated 36–37, 102
porter myosins 105–106
power stroke 7, 30, 40, 46–47, 95, 117, 123–124, 126, 128, 130–131, 136, 149, 157–159, 164, 167
  amplitude of 78, 116–118, 123–124

proteins, regulatory 30
protozoan animals 241–242

rabbit cardiac muscle myosins 40
rabbit psoas muscle 144
reactive lysine residue (RLR) 66, 82, 88–89, 97–98
rigor actin-myosin linkages 16
rigor actin–myosin complex 31
  static structure of 31
rigor actin–myosin linkages 104, 115
rigor fibers, low-Ca 96–97
rigor linkages 115, 118, 125, 128
rigor muscle 20, 132, 134
rigor muscle fibers 16–17, 20, 59
  frozen 18
  skinned 21
  stretch of 16, 20
  stretched 23
rigor state 16, 19, 31, 50–53, 55, 57–58, 94–95, 132, 134, 157
  low-Ca 96
RLR, *see* reactive lysine residue

sarcomere elements 23
sarcomere length 23, 25, 50, 115, 144, 176–177, 184–188, 205, 234–235
sarcomere shortening 23, 207–208, 227, 234–235, 237–238
sarcomeres 5–7, 14, 16–17, 24–25, 51, 54, 115, 148, 167, 187–188, 215–216, 228–232, 234–236, 239
sarcoplasmic reticulum (SR) 135, 140, 208–209, 218–219, 234
shortening velocity 9, 86, 90, 92, 104, 108–109, 121, 142, 145–147, 159–160, 165, 235
  sarcomere 235
skeletal muscle 1–5, 7–9, 28, 43, 50, 65, 72, 141–143, 155–156, 172, 179–180, 191
skeletal muscle fibers 187, 211–212, 238
skeletal muscle myofibril 150
skeletal muscle myosin 45, 68, 80
skinned fiber bundles 144
skinned fiber preparation 144
skinned fibers 16, 94, 101, 144, 168
  single 55
skinned muscle fibers 18, 21, 37, 51, 85, 94–95, 99, 110–111, 120, 136, 143–145, 157
  myofilament lattice structure of 110
slack test 108–109
sliding filament mechanism 3, 7–8, 174
slime mold 44–45
smooth muscle 68
spindle-shaped filaments, synthetic 74
SR, *see* sarcoplasmic reticulum
stretched muscle fibers, enhancement of mechanical performance in 174–175
striated muscle 1–2, 6

swinging lever arm hypothesis 43, 45–47, 61, 83, 123–125
swinging lever arm mechanism 43–44, 46, 83–84, 104, 165
synthetic filaments 68–69, 72
velocity of 210, 216
transverse spread of local contraction 208
tropomyosin 7, 30, 95
twitch 134, 193–194
twitch-tetanus ratio 239

T tubule membrane 209, 211, 213, 215, 228, 235, 238–239
T tubule network 210, 216–217, 233
time-resolved X-ray diffraction 20, 178–179, 181–182, 185
transmutation theory 1
transverse muscle stiffness 134
transverse spread of contraction 203, 210, 224–225, 227–229, 231, 233, 235, 237, 239

ultrasonic waves 132–133
unloaded shortening 109, 165–167, 169

vertebrate muscle myosin filaments, synthetic 74–75, 77, 79, 81
vertebrate muscle myosin filaments of myosin head, synthetic 79